Introduction to

In Vitro
Cytotoxicology

Mechanisms and Methods

Introduction to
In Vitro
Cytotoxicology

Mechanisms and Methods

Frank A. Barile, Ph.D.

Department of Natural Sciences
Health Professions Division
City University of New York at York College
Jamaica, New York

CRC Press
Taylor & Francis Group
Boca Raton London New York

CRC Press is an imprint of the
Taylor & Francis Group, an **informa** business

CRC Press
Taylor & Francis Group
6000 Broken Sound Parkway NW, Suite 300
Boca Raton, FL 33487-2742

Reissued 2019 by CRC Press

ISBN 13: 978-0-367-22538-4 (hbk)
ISBN 13: 978-0-367-22540-7 (pbk)
ISBN 13: 978-0-429-27548-7 (ebk)

Visit the Taylor & Francis Web site at http://www.taylorandfrancis.com and the
CRC Press Web site at http://www.crcpress.com

PREFACE

Cell culture technology, and accompanying *in vitro* cytotoxicology, is an important, newly developing discipline of modern toxicology and is gaining increasing acceptance in the field. This acceptance has been accomplished in part through the inevitable turn of events in society which have forced research scientists and regulatory toxicologists to assume more ethical and humane approaches to toxicity testing. Public demands have compelled administrators and legislators to impose stricter regulations governing the use of animals in biomedical sciences. As a result, an increasing number of applied toxicologists, such as those with traditional training in cell biology, physiology, analytical chemistry and biochemistry, have adapted their animal toxicity testing protocols to *in vitro* techniques. These procedures are having a profound influence in toxicity testing and cannot be ignored, as has been the tendency in the past. Because of the notoriety that *in vitro* cytotoxicity testing has been enjoying, a basic knowledge of cell culture and cytotoxicity testing methods in toxicology courses taken by health professional and biomedical students, including those encountered in schools of toxicology, environmental health, pharmacy, and human, dental, and veterinary medicine, is necessary. Consequently, this text is useful as part of general and upper level toxicology courses which incorporate cell culture methodology, mechanisms of cytotoxicity, and procedures used in toxicity testing. In addition, the text serves as a useful tool for academic, industrial, and regulatory toxicologists, cell biologists, pharmacologists, and animal welfare organizations.

In general, the text introduces *in vitro* cytotoxicology for the benefit of toxicologists and other professionals. Some misunderstandings and concerns regarding *in vitro* toxicity assays will be reduced and the potential value of the model systems will be appreciated as alternative methodologies develop. There are many toxicologists today who consider cell tests for general toxicity to be immature analogues to the mechanistic short-term genotoxicity tests. In reality, the general toxicity cell tests are neither typically short-term nor mechanistic, but more often are empirically based quantitative tests of cell injury due to a host of various undefined mechanisms. In addition, the explanations developed to understand certain toxic phenomena are usually more precise when cell tests are employed than with the traditional animal experiments. Consequently, the text presents the prevailing mechanisms which underlie *in vitro* cytotoxicology and cellular toxicity.

The chapter on cell culture methods is a useful reference for students and toxicologists who wish to become familiar with the terminology and general principles. It can also serve as a guide for researchers interested in establishing a cell culture laboratory. The background, techniques, and use of cellular tests are described in detail in some areas where the procedures are widely used or have withstood the test of time and are only summarized in other areas. This serves to introduce certain well-known assays, while detailed descriptions of procedures which are currently in use but not universally accepted may not be practical. The methods are presented to the fledgling investigator as a starting point and general guide for further investigation, rather than as a detailed flowchart for particular procedures. Thus, the information furthers understanding of the principles of cytotoxicity.

The ability of the tests for predicting or screening for human toxicity is presented according to the current advantages and limitations which formulate the scientific and regulatory basis for human risk assessment. Thus the optimum utilization of certain tests for specific groups of chemicals is discussed in light of the validation programs in progress. As a result, continuous evaluation of assays for general and organ-specific toxicity, as well as tests for genotoxicity and carcinogenicity, is currently being investigated by several toxicology groups throughout the world. The results of these efforts are anxiously awaited by the biomedical and toxicological communities, and by public interest groups.

Frank A. Barile, Ph.D.

THE AUTHOR

Frank A. Barile, Ph.D., is Associate Professor in the Health Professions Division, Department of Natural Sciences, at the City University of New York at York College.

Dr. Barile received his B.S. in Pharmaceutical Sciences in 1977, his M.S. in Pharmacology in 1980, and his Ph.D. in Toxicology in 1982 from the College of Pharmacy at St. John's University, New York. After doing a post-doctoral fellowship in Pulmonary Pediatrics at the Albert Einstein College of Medicine, Bronx, New York, Dr. Barile moved to the Department of Pathology, Columbia University, St. Lukes/Roosevelt Hospital, New York, as a Research Associate. In these positions he investigated the role of pulmonary toxicants on collagen metabolism in cultured lung cells. In 1984, he was appointed as Assistant Professor in the Department of Natural Sciences at CUNY and became Associate Professor in 1991.

Dr. Barile holds memberships in several professional associations, including the American Association for the Advancement of Science, Tissue Culture Association, Sigma Xi National Scientific Honor Society, Rho Chi National Pharmaceutical Honor Society, New York Academy of Sciences, Scandinavian Society of Cell Toxicologists, American Association of University Professors, Italian-American Institute of CUNY, and the CUNY Academy of Sciences. He has been appointed as a Consultant Scientist with several clinical and industrial groups, including the Department of Pediatrics, Pulmonary Division, of the Long Island Jewish Medical Center in New Hyde Park, New York, and Pharmacon International, New York.

Dr. Barile has been the recipient of research grants from the National Institutes of Health, including awards from the Minority Biomedical Research Support (MBRS) program, the Minority High School Student Research Apprentice (MHSSRAP) program, and the Research Foundation of CUNY.

Dr. Barile has authored and co-authored approximately 50 papers in various peer-reviewed biomedical and toxicology journals along with distinguished international investigators. He also lectures regularly to undergraduate students in the health professions in toxicology and biomedically related courses. Dr. Barile continues fundamental research on the cytotoxic effects of therapeutic drugs and environmental chemicals in cultured mammalian lung cells.

CONTRIBUTORS

Thomas W. Sawyer, Ph.D., D.A.B.T.
Research Toxicologist
Biomedical Defence Section
Defence Research Establishment Suffield
Medicine Hat
Alberta, Canada

Bjorn Ekwall, M.D., Ph.D.
Assistant Professor
Department of Toxicology
University of Uppsala
Biomedical Center
Uppsala, Sweden

Dedicated to my parents, Mr. and Mrs. Frank Barile, Sr., for their encouragement, advice, and support throughout the years.

ACKNOWLEDGMENTS

Appreciation is extended to the National Institute of General Medical Sciences (NIGMS), National Institutes of Health (U.S.), for the years of grant support for the research performed in my laboratory.

Several individuals and colleagues, whose interests, efforts, and concerns for the development of cell culture models to replace animals in toxicity testing, have enabled me to carry my work to completion and to make this book a reality. Among them, words of thanks are in order to Professors Robert Macholow and Hope Young, Ms. Myra James, Ms. Lolita Borges, and to my children, Frank, Lauren, and Matthew.

Finally, my thanks are extended to my friends and colleagues: Dr. Thomas Sawyer, for his contribution to the project, and Dr. Bjorn Ekwall, without whose extensive discussions of the fundamentals of cytotoxicology and encouragement over the years, this book may not have materialized.

CONTENTS

Chapter 1
Cell Culture Methodology and Its Application to *In Vitro* Cytotoxicology

Chapter 2
Mechanisms of Cytotoxicology

Chapter 8
Pharmacokinetic Studies in Cellular Systems

Chapter 9
Cellular Methods of Genotoxicity and Carcinogenicity
Thomas W. Sawyer

Chapter 10
Experimental Design and Statistics

Chapter 11
Standardization and Validation
Bjorn Ekwall and Frank A. Barile

Chapter 12
Cell Culture or Animal Toxicity Tests? Or Both?

Chapter 1

CELL CULTURE METHODOLOGY AND ITS APPLICATION TO *IN VITRO* CYTOTOXICOLOGY

I. INTRODUCTION

A. *IN VITRO* TOXICOLOGY

The techniques to grow animal and human cells and tissues on plastic surfaces or in suspension have contributed significantly to the development of biomedical science. Thus the term *in vitro* refers primarily to the handling of cells and tissues outside of the body under conditions which support their growth, differentiation, and stability. Cell culture techniques (and for the purposes of this book, cell culture and tissue culture are used interchangeably) are fundamental for understanding and performing the procedures in cellular and molecular biology and have been used with increasing frequency in medicinal chemistry, pharmacology, genetics, reproductive biology, and oncology. Cell culture technology has improved throughout the decades, partly as a result of the interest in the methods rather than through its applications. Much of the mystery originally surrounding the techniques has waned, but the basic principles underlying the establishment of the methodology remain. That is, tissue culture represents a tool where scientific questions in the biomedical sciences can be answered through the use of isolated cells or tissues, without the influence of other organ systems. With this understanding, the use of isolated cells does not purport to represent the whole human organism, but can contribute to our understanding of the workings of its components.

B. TEST SYSTEMS

The use of cell culture techniques in toxicological investigations is referred to as *in vitro cytotoxicology*, or *in vitro toxicology*, the latter term including non-cellular test systems such as isolated organelles. The reasons for the popularization of these techniques in toxicology are numerous:

1. Since the growth of the first mammalian cells were described in capillary glass tubes, the technology has progressed and has been extensively refined.
2. Based on *in vitro* cytotoxicological methods, the direct interaction of chemicals with human and animal cells *in vivo* are responsible for the toxic effects demonstrated.
3. The necessity to determine the toxic effects of industrial chemicals and pharmaceuticals that are developed and marketed at rapid and unprecedented rates has provoked the need for fast, simple, and effective test systems.

4.   Although the normal rate of progression of any scientific discipline is
      determined by the progress within the scientific community, some areas
      have received more encouragement than others.

Specifically, the recent arguments and open protests of animal rights activists have
forced researchers and regulatory agencies to direct research initiatives toward the
development of alternative methods of toxicity testing.

   To understand the possibilities for, and more importantly, the limitations of
cell culture methods in toxicology, it is necessary to be acquainted with the
main features of the techniques to culture cells and tissues. Unlike other types
of biomedical research, continuous care of the cells in culture is required for
extended periods, which necessitates planning.

II. HISTORY

A. TISSUE EXPLANTS

   In the fledgling years of cell culture, U.S. scientists removed tissue explants
from animals and allowed them to adhere to glass cover slips or placed them
in capillary tubes in clots formed from lymph or plasma. It was discovered
early on that the tissue or organ, which was once an intact biological specimen,
would exhibit a breakdown in the supporting matrix with the migration of
individual cells from the main body of the specimen as a consequence.[1] With
the addition of serum or whole blood, the resulting clot formed a hanging drop,
making it possible to look at these cells through an ordinary light microscope.
With some refinement, the explants, with outgrowing cells, were cultivated in
small glass flasks in plasma or embryonal extract. Cells were transferred from
one flask to another by scraping and loosening the cells with a rubber "policeman". In the 1940s synthetic media were developed (Earle, Parker, Eagle)
which were used together with various serum additives. The main problem
encountered was bacterial contamination which, because of the more rapid rate
of mitosis, usually outpaced the growth of the mammalian cells. This contamination generally resulted in overwhelming bacterial cell growth and disintegra-
tion of the cell cultures. In the 1950s, this setback was largely overcome by the
addition of liquid antibiotics to the media.[2] Later on, the development of better
aseptic techniques, such as the incorporation of sterile, disposable glassware,
autoclave units, and laminar air flow hoods, made antibiotics superfluous in
most cases. With an increase in the understanding of the influence of pH,
buffers, and ambient environment, and the incorporation of chemically inert
plastics and microprocessor controlled incubators came the realization of the
full potential of *in vitro* technology.

B. RECENT DEVELOPMENTS

   Today, cell biologists have further developed culture techniques to aid in the
understanding of cellular/extracellular interactions, such as mesenchymal-
epithelial relationships and epithelial-cell matrix interactions. These *in*

vitro studies have materialized largely through the development of additions to accepted protocols, such as chemically defined cell culture media, addition of cellular substrata to culture flasks, porous membranes which allow for the passage of low molecular weight soluble materials, treated plastic surfaces, and incubation of cells with cocultures.

III.   EQUIPMENT

In the simplest case, cell cultivation only requires a sterile air flow work station, a temperature-controlled incubator, an autoclave, a hemacytometer, custom gas tanks, a source of ultrapure water, and pipettors. Biological safety cabinets will reduce bacterial and fungal contamination, and will also protect the operator from exposure (Figure 1). Automatic glassware washing facilities, an electronic cell counter, an incubator which monitors humidity and environment, a rate-controlled peristaltic pump, and camera equipment will facilitate experiments and allow for less troublesome, routine operations (Figure 2). The establishment of tissue culture laboratories as part of toxicological investigations has taken on a more sophisticated and dedicated role. In general, for reasons of maintaining sterility, viability, and identification of cultured cells, most tissue culture laboratories are dedicated to this function. However, complete separation of all functions is usually not necessary. For instance, an analytical balance, pH meter, hot plate, and a magnetic stirrer can be used in other laboratory areas. Media preparation equipment and glassware are generally reserved for the cell culture lab.

Most supplies and plasticware used in handling cell cultures on a daily basis are sterile and disposable. These include borosilicate glass pipettes, tissue culture flasks and bottles, petri dishes, polypropylene and polyethylene centrifuge tubes, and filter units.

IV.   THE CULTURED CELLS

A. PRIMARY CULTURES

It is easier to grow cells from embryonal tissue than from adult donors. In general, the younger the donor, the more replications and faster replication time can be expected from the resulting cells. Some cells derived from neoplastic tissue are, in this respect, similar to embryonal cells. Ideally, cell cultivation begins with the aseptic removal of an explant ($1/2 \times 1$ mm) from an animal or human organ. The cells of the explant are mechanically or enzymatically separated from the matrix and are allowed to grow in medium in contact with the bottom surface of the culture vessel. While the central core of the explant often atrophies due to less favorable diffusion of nutrients from the medium, the peripheral cells migrate and multiply. This establishes a *primary culture* where two parallel processes occur: (1) The differentiated cells of the original

Figure 1. Photograph of biological safety cabinet used for tissue culture experiments. Most cabinets are equipped with HEPA filters which provide sterile air flow as well as operator protection from hazardous materials. (Courtesy of Labconco Inc.)

explant (if there were any) usually do not divide, and with time, will successively lose some of their specialized functions (dedifferentiation); and (2) less specialized cells, e.g., fibroblasts, divide rapidly, and will eventually outnumber the specialized cells.

B. FINITE CELL LINES

If the primary culture is transferred to a new culture vessel with new, fresh medium, it is designated as a *cell line*. The subcultivation (passage) is made by detaching the cells from the glass or plastic using chemical (Versene, a calcium chelator) or enzymatic (trypsin) methods and then the cells are dispersed and inoculated in a number of new vessels (Figure 3).[3] Because of selective survival of viable cells, the cell line will be more homogenous and dedifferentiated with time. A culture which is demonstrating adequate growth is harvested every 3 to 7 days and subdivided to several new flasks, indicating a doubling rate of 48 to 72 hours. After 40 to 50 cell divisions (also referred to as population doubling level and may require months in culture depending on growth rate), most cell lines stop growing and ultimately die, possibly due to the timing of the genetic program (Figure 4). Such a *finite cell line* has a constant number of chromosomes (diploid number) and exhibits an orderly

Figure 2a. Photograph of single and dual chamber water-jacketed CO_2 incubators, with electronic temperature control and safety alarm systems with digital display.

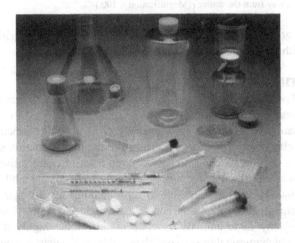

Figure 2b. Photograph of disposable plastic and glass and reusable glass supplies used in routine cell culture laboratories (from Corning® Corp., Corning, NY). For cell culture, the disposable plastic apparatus includes multiple well plates, roller bottles, tissue culture flasks and plates, and plastic tubes. Disposable pipettes, cryogenic vials, centrifuge tubes, filtration units, and vacuum filter systems are available for liquid handling, storage, and separation.

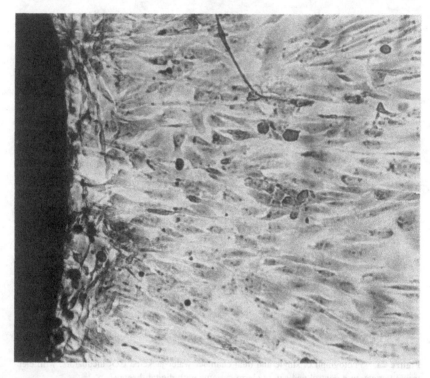

Figure 3.   Phase contrast micrograph of epithelial cells emerging from explanted lung tissue in culture. Note the dark contrast of the explant tissue and the change in phenotypic appearance as cells migrate away from the source. (Magnification × 100.)

orientated growth pattern, including inhibition of growth of individual cells by contact with its neighbor cells (Figure 5a and 5b, Figure 6).

C. CONTINUOUS CELL LINES

A small percentage of cell lines will not die out with time, but are transformed to ***continuous cell lines***, with a growth pattern often referred to as immortal. Transformation occurs either spontaneously or as a result of incubation with viruses or chemicals. The exact mechanism of the transformation is still undefined, but a continuous cell line has acquired a set of characteristics, such as varied chromosome number (heteroploidy), and loss of contact inhibition. Moreover, continuous cell lines are able to form colonies in soft agar media, i.e., without help of glass or plastic contact, and induce tumors if implanted in immunologically nude animals. In spite of this, some highly differentiated functions, which mimic those of specialized cells, will persist in continuous cell lines. These differentiated functional markers are used to characterize the line and are relied upon to monitor the progress of the cells in culture.[4-5] They are also often used as the parameters for assessing *in vitro* toxicity. Table 1 presents some cell lines which are often used in toxicological studies.

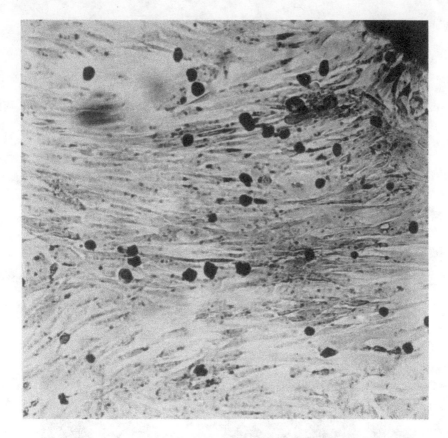

Figure 4. Phase contrast micrograph of fibroblasts emerging from explanted lung tissue in culture. Note their spindle shaped appearance, with rounded cells sparsely populating the culture. (Magnification × 100.)

D. CELL STRAINS

It is possible to select a single cell in a culture and subcultivate this cell to form a *clonal cell line*. Cloning allows for the selection of one type of cell in a culture (e.g., cells with specific functional markers) and the establishment of a subculture in which the desired characteristic has been passed on to the progeny (identified as a *cell strain*).

E. CLONAL GROWTH AND MAINTENANCE CULTURES

By taking specific measures, the highly differentiated parenchymal (epithelial) cells of the explant can be maintained in primary culture. Such measures include the use of serum-free medium, arginine-free medium, coating of the plastic culture surfaces with cell-specific matrix (laminin or fibronectin) which promotes growth and adhesion of either epithelial or mesenchymal cells, respectively, and/or separation of the cell types by clonal growth or *differential*

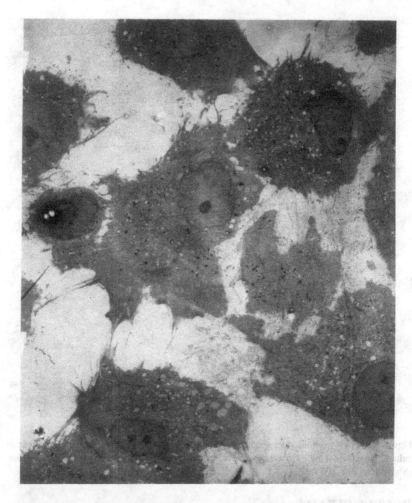

Figure 5a.   Electron micrograph of cultured lung epithelial cells. Dense electron opaque granules, corresponding to phospholipids, and numerous microvilli, characterize this cell type. The
presence of condensed nuclei and limited amount of heterochromatin suggests mitotically active
cells. (Magnification × 3630.)

adhesion. The last technique separates epithelial cells from fibroblasts based
on the ability of the latter cells to settle and adhere to the surface more quickly.
One to three hours after an initial inoculation of the culture with the cell
suspension, the medium is removed and the epithelial cells, which have not
adhered, can be cultured in a separate vessel. The primary culture of the
specialized cells now serves as a *maintenance culture*.

Normal primary cultures of specialized cells, including those derived from
adult liver, may be kept for weeks, while cells derived from embryonal liver,
neuronal, or lung epithelial cells are viable for months. There is always a
tendency of the established culture to lose specific functions with time, but this
process may be counteracted by culturing the specialized cells on a feeder layer

Figure 5b. Electron micrograph of cultured lung epithelial cells. The formation of junctional complexes along the length of their cytoplasmic borders is a hallmark of these cells. (Magnification × 23,100.)

of other cell types or by using serum-free medium. Such measures sometimes may even induce specialized functions in less differentiated epithelial cells, including cell lines. Isolation of primary cultures, however, requires repeated use of animals for the extraction and establishment of explants, which entails expense and is labor intensive.

F. CRITERIA FOR IDENTIFICATION

Several criteria are used for the identification and classification of cells in culture. These methods are employed when the cell line is established and periodically to monitor the genetic purity of the cells.[6] The criteria include:

1. *Karyotypic Analysis* — For a cell line designated as being derived from normal tissue, the chromosome complement should be identical to the parent cell or the species of origin. In this way, the cell line may be classified as diploid. Any other designation would classify the cells as

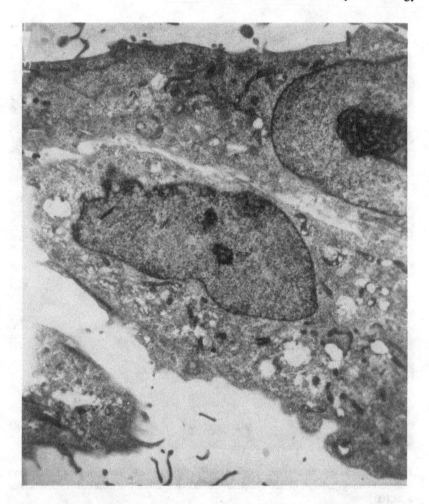

Figure 6. Electron micrograph of cultured lung fibroblasts. The spindle shaped morphology, granular cytoplasm, and limited amount of rough endoplasmic reticulum, are characteristic of these cells in culture. (Magnification × 9900.)

aneuploid or heteroploid. Continuous cell lines derived from tumors, or transformed cells, become part of the latter categories.

2. *Aging in Culture* — The life span of cells in culture is measured according to the population doubling level (pdl). This criterium refers to the number of times a cell undergoes mitosis since its original isolation *in vitro*. The general formula for assessing pdl is:

$$pdl_f = 3.32 \, (\log F - \log I) + pdl_i$$

where pdl_f = the final pdl at the time of trypsinization or at the end of a given subculture; F = the final cell count; I = the initial cell number used

TABLE 1
Some Commonly Used Cell Lines

Origin	Name	Notes
Human, cervix uteri	HeLa	Continuous
Human, adult lung, epithelial	A549	Continuous
Human, trachea	HEP 2	Continuous
Human, liver	Chang liver	Continuous[a]
Human, epidermoid tumor	KB	Continuous
Human, embryonal lung, fibroblast	HFL1, WI-38, IMR-90, MRC-5	Finite
Mouse, embryo	3T3	Continuous
Mouse, lymphoma	L5178	Continuous
Mouse, neuroblastoma	C1300	Continuous[a]
Rat, intestine	IEC-17	Finite
Rat, lung epithelial	L2	Continuous
Chinese hamster, ovary	CHO	Continuous
Hamster, kidney	BHK 21	Continuous

[a] Several functional markers persist.

as inoculum at the beginning of the subculture; pdl_i = the doubling level of the inoculum used to initiate the subculture.

Cells with finite life spans generally show signs of aging, such as loss of cell shape and increase in the cytoplasmic lipid content. The number of pdl that a particular finite cell line exhibits depends on the age of the organ of origin. Permanent cell lines are capable of indefinite continuous multiplication *in vitro*, provided they are maintained under optimum conditions.

3. *Anchorage-Dependent Cultures* — Some cells are capable of growing in suspension culture, while others require attachment to a matrix. This matrix may take the form of polyethylene plastic, treated so that its electrostatic properties allow isolated cells to attach and migrate. Other attachment matrices include components of the extracellular matrix, such as laminin and fibronectin. These components mimic the *in vitro* attachment substrates normally present in the epithelial lining and the interstitium, respectively. Thus, they may be selective for cells of epithelial or mesenchymal origin. Still other support matrices use microporous filter membranes which not only allow for cellular attachment but also permit passage of macromolecules through the membranes to the basolateral surface of the cell layer.

4. *Contact Inhibition* — Cells which exhibit contact inhibition will arrest their cell cycle in the G_0 phase when the layer of cells completely occupies the surface available for proliferation, thus forming monolayer cultures. Cells which do not exhibit contact inhibition will grow in multilayers.

Other methods of cell classification are not necessarily routinely performed. These include analysis of tissue-specific differentiated properties, such as secretion of macromolecules or the presence of intracellular enzymes, ultrastructural identification, fluorescent labeling for identifiable structural proteins, or the ability of the cells to form an invasive malignant tumor when injected into immunologically deficient (*nude*) mice.

V. MEDIA

A. CHEMICALLY DEFINED MEDIA

Cells are often cultured in a chemically defined medium after the addition of fetal or newborn calf serum. The medium is composed of a buffered saline solution containing soluble amino acids, carbohydrates, vitamins, minerals, and cofactors. Optional ingredients include a pH indicator, separate buffering systems, such as HEPES buffer, and some non-essential amino acids which may be used by particular cell types. The formulas for these media are listed in any company catalog which supplies the readily soluble powdered or liquid media. Some of the more commonly used media are modified Eagle's medium (MEM), basal medium Eagle (BSE), Dulbecco's modified Eagle's medium (DMEM, Table 2), and Ham's F12 medium. These media are generally designed for use with serum or serum proteins. Serum is a complex mixture of poorly defined biological components consisting of cell growth factors, transport proteins, hormones, trace metals, lipids, and biomatrix adhesion factors. Without the addition of a minimum percentage of serum (5 to 20%) as part of the growth formulation, cellular proliferation does not proceed favorably.

In cell culture, balanced salt solutions function as washing, irrigating, transporting, and diluting fluids while maintaining intra- and extracellular osmotic balance. In addition, they provide cells with water and bulk inorganic ions essential for normal cell metabolism and are formulated to provide a buffering system to maintain the medium within the physiological pH range (7.2 to 7.6). The solutions may also be combined with a carbohydrate, such as glucose, thus supplying an energy source for cell metabolism. The most commonly used prepared salt solutions include Dulbecco's phosphate buffered saline, Earle's balanced salts, and Hank's balanced salt solution (HBSS, Table 3).

In addition to satisfying nutrient requirements, the cells must be placed in an environment which mimics the *in vivo* situation. Table 4 summarizes the conditions which support survival and multiplication, and include ***temperature, pH, carbon dioxide tension, buffering, osmolality, and humidification***. These parameters vary somewhat with the cell type, but are very similar when growing various mammalian cell lines.

1. Temperature

For most mammalian cells, the optimum ***temperature*** for cell proliferation and differentiation is 37°C. Cells withstand falls in temperature more easily than rises in temperature. In fact, temperatures as low as 4°C may reduce

TABLE 2
Components of a Typical Cell Culture
Growth Medium, Dulbecco's Modified
Eagle's Medium

Component	1 × Liquid (mg/L)
Inorganic salts	
$CaCl_2$ (anhyd.)	200.00
$Fe(NO_3) \cdot 9H_2O$	0.10
KCl	400.00
$MgSO_4$ (anhyd.)	97.67
$MgSO_4 \cdot 7H_2O$	200.00
NaCl	6400.00
$NaHCO_3$	3700.00
$Na_2HPO_4 \cdot H_2O$	125.00
Other components	
D-Glucose	1,000.00
Phenol red	15.00
Sodium pyruvate	110.00
Amino acids	
L-Arginine HCl	84.00
L-Cystine	48.00
L-Cystine·2HCl	63.00
L-Glutamine	584.00
Glycine	30.00
L-Histidine HCl·H_2O	42.00
L-Isoleucine	105.00
L-Leucine	105.00
L-Lysine HCl	146.00
L-Methionine	30.00
L-Phenylalanine	66.00
L-Serine	42.00
L-Threonine	95.00
L-Tryptophan	16.00
L-Tyrosine	72.00
L-Tyrosine·2Na·$2H_2O$	104.00
L-Valine	94.00
Vitamins	
Ca pantothenate	4.00
Choline chloride	4.00
Folic acid	4.00
i-Inositol	7.20
Niacinamide	4.00
Pyridoxal HCl	4.00
Riboflavin	0.40
Thiamine HCl	4.00

Courtesy of Grand Island Biological Company
(GIBCO), Grand Island, NY.

TABLE 3
Components of a Typical Cell Culture
Physiological Salt Solution, Hank's Balanced
Salt Solution[7]

Component	6 mM	
	g/L	mM
Inorganic salts		
$CaCl_2$ (anhyd.)	0.14	1.3
KCl	0.40	5.0
KH_2PO_4	0.06	0.3
$MgCl_2 \cdot 6H_2O$	0.10	0.5
$MgSO_4 \cdot 7H_2O$	0.10	0.4
NaCl	8.00	138.0
$NaHCO_3$	0.35	4.0
Na_2HPO_4	0.048	0.3
Other components		
D-Glucose	1.00	5.6
Phenol red	0.01	0.03

Note: Formula for higher concentrations are also available.

Courtesy of Grand Island Biological Company Catalog (GIBCO),
Grand Island, NY.

metabolic activity, but will not irreversibly inhibit biological functions. Temperatures as high as 39°C may stimulate the formation of heat shock proteins resulting in irreversible ultrastructural and functional damage.

2. pH

The *pH* range that supports optimum cellular multiplication is 7.2 to 7.4, with some cell growth at the extremes of 7.0 and 7.6. Some cells are more tolerant of slightly acidic conditions than corresponding basic conditions, and transformed cells are often more resistant to changes in pH than diploid cells.

3. Carbon Dioxide Tension

Most culture media contain bicarbonate as part of the buffering system to prevent large and rapid changes in pH. At standard temperature and pressure in solution, bicarbonate and carbonic acid are in equilibrium as determined by the pH. At 37°C, however, carbonic acid equilibrates with gaseous *carbon dioxide*, thus driving the carbonic acid into the gaseous phase. This makes it impossible to maintain an adequate concentration of the acid and bicarbonate in the culture medium without also maintaining an increased partial pressure of carbon dioxide in the gas phase above the liquid. Balanced CO_2 tension and maintenance of pH is accomplished with modern water-jacketed CO_2 incubators, which adequately maintain the gaseous phase in the chamber at controlled levels. Therefore, most

TABLE 4
Conditions Necessary for Support and
Proliferation of Cultured Cells

Criteria	Parameter
Culture conditions	Sterility
	pH
	Buffered growth medium
	Buffered wash medium
	Temperature
	Humidity
	Osmolality
	Time
	Ultrapure water
In vitro	Fetal bovine serum
supplementation	Addition of physiological
	substrata to culture
	vessels
	Culture vessel inserts
	Defined culture additives
	Pretreated plastic flasks
	(Primaria®)
	Antibiotics
	Co-cultures
Optional additives	pH indicators
	HEPES buffer

culture media are formulated for use with 2 to 10% CO_2, so that the appropriate pH can be maintained throughout the life cycle of the cells.

4. Buffering

When petri dishes or multiwell plates are removed from the incubation chamber, carbon dioxide necessarily escapes. This inevitably results in the inability of the pH to be adequately maintained in the absence of the gaseous phase above the liquid. This occurrence is not usually a problem when the cultures are removed during washing steps with physiological salt solutions or replacing the expended medium. When *in vitro* experiments necessitate long periods outside of the incubation chamber, however, a buffering system is required. Many laboratories now routinely incorporate organic *buffers*, such as HEPES, in the media formulations to prevent the rapid shift of pH when cultures are removed from the CO_2 incubator. Sufficient sodium hydroxide or sodium bicarbonate is added to reach the working pH. Alternatively, cells are grown in tissue culture flasks, which can be gassed with gas mixtures separate from the incubator, and specially formulated for the medium. The individual flasks are gassed and sealed with screw caps, thus preventing escape of the gaseous phase. Although this method is tedious, it may also be helpful when volatile organic liquids or gases are tested in toxicity testing protocols.

5. Osmolality

The optimum range of *osmolality* in the culture medium depends on the cell type, but it is generally between 250 and 325 milliosmoles per kilogram (mOsm/kg). The osmolality of the solution is established during the preparation of the powdered medium, and is standardized when the product is obtained. The concerns arise during the period of incubation when the relative *humidity* in the chamber is kept at near saturation. When the incubator door is opened, dry air enters and cools the chamber, thus releasing the humidified air by evaporation. An incubator which does not provide for adequate heat transfer will remain cool. If the liquid media in the cultures is at a higher temperature, the water in the media will evaporate, leaving behind a medium solution with a higher osmolality. To prevent evaporation and to maintain the osmolality of the culture medium, rapid equilibration of heat throughout the chamber is necessary. Most modern water-jacketed incubators are equipped with internal fans which distribute the heat. These instruments are also precisely insulated so as to prevent the formation of cool spots, which prevent the humidification of the chamber. In general, the chamber humidity is maintained by leaving an open trough of water, preferably with a large surface area. Adequate humidification is determined visually by the presence of condensation on the glass door of the incubator, but no condensation should be formed on the culture dishes. Alternatively, humidification is monitored by measuring the osmolality of control media incubated parallel to cell cultures. If the osmolality is too high, as compared with fresh medium at the start of a cell cycle, then the humidity is not enough and evaporation is occurring.

6. Water Requirement

A fundamental requirement of cell culture systems, and one that is often overlooked in the initial steps of establishing the cell culture laboratory, is the quality of the water used to make the media and salt solutions. In general, distilled water or a single passage of water through a deionizing column is not sufficient to remove all of the impurities which remain from the feed water. Contaminants in the water may take the form of trace metals, organics, variable amounts of divalent cations such as magnesium or calcium, and metabolic products of microorganisms. Such contaminants will interfere with cell growth and metabolic processes. These impurities are essentially removed with systems that recirculate water through a series of deionizing and organic exchange columns. Most of these systems have a digital display which shows the purity of the collected water in milliosmoles. At present, water resistivity measuring at least 10 megaohms is suitable for culturing cells, and the limit of purification with many of the exchange columns is about 18 megaohms. In addition, some water purifying systems can also yield water which is sterilized by gamma irradiation at the collecting end. Several water purifying systems are pictured in Figure 7.

7. Glassware Maintenance

Another aspect of maintaining a cell culture laboratory at optimum efficiency is the manner in which the glassware is handled, washed, and stored. All glassware designated for use as part of the cell culture area should not be used for storage of other chemicals or for experiments not related to the growth of cells. Glassware should be clearly marked *for cell or tissue culture only* and should be washed in detergent that is compatible with the objectives of the lab. Essentially, the detergent should leave no residue adhering to the pores of the glass, and should be easily rinsed with deionized water. The glassware should also be stored separately from the rest of the equipment and materials kept in the research laboratory.

B. SERUM-FREE MEDIA

A trend in modern cell culture has been to replace serum by a mixture of synthetic or otherwise defined factors, such as hormones (insulin, transferrin), trace metals (selenium, manganese), growth factors (epidermal growth factor, fibroblast growth factor), and extracellular matrix components (laminin, fibronectin). These selective *serum-free* media facilitate adaptation of cells to the culture and allow for better standardization of experimental conditions. Serum is a complex and poorly characterized mixture, the composition of which may vary according to the commercial batch; the concentrations of some components essential for cell growth may vary. Serum may contain naturally occurring substances or microbiological contaminants (e.g., mycoplasma, viruses, endotoxins) that are toxic for certain types of cultures.[10,11]

Serum-free medium facilitates the isolation of the desired cell type. When primary cell cultures are established, the use of serum-free medium virtually eliminates the overgrowth of fibroblasts that usually grow rapidly in serum-supplemented media. The advantages of serum-free (or low serum) media have been demonstrated in the establishment of differentiated rodent thyroid cells in hormone-supplemented medium containing only very small amounts of serum.[8] Other cell lines form differentiated features which are not necessarily expressed in serum-supplemented media. For example, the MC845 line forms villus-like secretory structures,[9] the HLE 222 human lung epidermoid carcinoma cells produce extensive keratinization,[11] and rat granulosa cells synthesize large amount of progestins and estrogens after stimulation by follicle-stimulating hormone (FSH).[12] Additional examples are reported by Barnes and Sato.[5,10]

Serum-free media must be individually prepared and adapted to each cell type. Such media have been described for use in maintenance cultures of many specialized cells, as well as some finite and continuous cell lines, such as human fetal lung fibroblasts and Chinese hamster ovary cells (WI-38, CHO,

a

b

Figure 7. Photograph of water purifying systems. (a) Water Prodigy™ benchtop model (Courtesy of Labconco Corporation, Kansas City, MO) contains a pretreatment cartridge, reverse osmosis membrane, granular activated carbon organic adsorption cartridge, mixed bed deionization cartridge and submicron and composite vent filters; (b) DF® System (Corning, Corning NY) contains a 5-μm prefilter, two deionization cartridges and a 0.2-μm final filter; (c) MP-190 Water Purification system™ wall mounted model (Corning) contains pretreatment, dual bed deionizing and final filtration cartridges and can deliver up to 90 liters/hour of Type I reagent water.

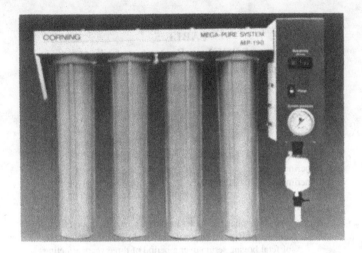

Figure 7c.

lung epithelial cells). Thus, optimization of growth using serum free defined media reduces experimental variation which may be encountered using different batches of sera and also encourages standardization of test procedures. In experiments conducted in our lab (Table 5), different lots of serum used over the course of 3 years accounted for varying rates of collagen production in human fetal lung cells, HFL1.

VI. PROCEDURES

A. STATIC AND PERFUSION CELL SYSTEMS

In general, culture conditions during an experiment or assay differ from those used during routine growth and maintenance of stock cultures. All stock cultures, and almost all experiments, are *static cell systems* which entertain a change of cell medium before or during subcultivation. During normal incubation, depletion of nutrients and accumulation of excretion products occur in the medium, resulting in less favorable cell culture conditions. Cell types that are anchorage dependent are often cultured as monolayers on the bottom surface of a flask covered with medium, in addition to an overlying gas phase consisting of 5 to 20% carbon dioxide in air. Cells may also be anchored on millipore filters or polystyrene particles. Rotating flasks or tubes are used to increase the culture area by providing contact of the sides and tops of the vessel with media, which also permits better oxygenation of cells. Some cell types (mainly continuous cell lines) need not adhere to a substratum and may be cultured in magnetically stirred suspension cultures or roller bottles. In addition, perfusion cultures are more advanced compared to static cell cultures since they allow for uniform flow of fresh medium by continuous irrigation of the flask. In addition to the usual cell culture manipulations, this system also uses positive pressure

TABLE 5
Effect of PGE1 on Percentage Collagen Production in Human Fetal Lung Fibroblasts: Influence of Different Lots of Sera[a]

	Control (%)	PGE1 (%)	P[b]
Experiment 1	6.6 ± 0.6	3.2 ± 1.2	<0.05
Experiment 2	6.1 ± 0.6	3.4 ± 0.8	<0.05
Experiment 3	10.0 ± 2.0	5.0 ± 1.0	<0.05
Experiment 4	12.2 ± 1.2	6.2 ± 0.2	<0.05

Note: The variation in the values for percentage collagen synthesis among the experiments reflects the use of different batches of fetal bovine serum over a period of three years affecting differences in expression of the collagen phenotype.

[a] [^{3}H]proline was used as the isotopic label. Percentage collagen synthesis calculated according to Barile et al.[14] Each value is the mean ± 1 SD for three separate cultures.

[b] P values for differences between control and PGE1 values were calculated using the two-tailed t test.

Reproduced from Barile, F. A., Ripley-Rouzier, C., Siddiqi, Z-e-A., and Bienkowski, R. S., *Arch. Biochem. Biophys.*, 265, 441, 1988. With permission.

pumps and micron filters to provide the cultures with oxygenated, waste-free medium. At present these systems are not routinely used in most laboratories. Apparently, they will require further testing to determine whether the more cumbersome and expensive instrumentation poses a distinct advantage over the static systems. It should be noted that the majority of cultures and test systems are much less oxygenated than animal tissues, and therefore permit anaerobic cell metabolism predominantly.

B. DISPERSION OF TISSUES

Many methods exist for the establishment of cell suspensions, and these methods largely depend on the cell type desired and the experience of the investigator who has optimized the isolation procedure.[3-5] Most of the commonly used procedures for isolating cell strains or cell lines, however, employ a variety of the same principles underlying tissue dispersion. That is, tissue can be disrupted and the cells dispersed mechanically, chemically, and/or enzymatically. In general, a typical protocol will use all of these manipulations. Mechanically, a tissue may be cut, minced, or sheared and may be passed through nylon mesh or sterile gauze. Additionally, the initial dispersion of tissue may be further disrupted with repetitive pipetting, magnetic stirring, or gentle vortexing. Chemical dispersion involves the removal of calcium from the dispersant solution or the addition of chelators, such as ethylenediamine tetraacetic acid (EDTA). Enzymatic dispersion requires the use of trypsin, chymotrypsin, collagenase, deoxyribonuclease, or pronase. Figure 8 shows a typical protocol for isolating cells from rabbit lung, outlining the general procurement, preparation, mechanical treatment of the tissue, dispersal in an enzymatic solution, and final recovery and plating of the suspended cells in growth medium. Although the method appears to be specific for this organ, this procedure is typical for isolating cells from intact organs or when biopsy specimens are available. Other procedures for specific organs are detailed in later chapters.

C. SUBCULTURING

Subculturing or passaging of cells involves the mechanical or enzymatic disruption of the cell monolayer and dividing and transferring them to new culture vessels. This procedure is usually required when the cell population has occupied most or all of the available surface area in the vessel. At this population density, normal diploid cells exhibit **contact inhibition**, i.e., cell growth slows or ceases as they come in contact with each other. At this semiconfluent or confluent stage, the cells are subcultured to allow for continued growth and proliferation.

Several methods exist for the dispersion of cells, including mechanical and enzymatic disruption. Mechanical methods involve the use of silicone or rubber cell scrapers which dislodge the monolayer in clumps. This method is not satisfactory for maintaining continued cultures since the process often compresses or lyses the cells. The method however is often used when terminating experiments for preparatory analyses. Enzymatic disruption uses proteolytic enzymes such as trypsin or collagenase, which yields more satisfactory cell numbers and affords less cell destruction. The enzyme solution is formulated as 0.5 to 2.0% in Ca^{+2}–Mg^{+2} free balanced salt solution, such as Dulbecco's phosphate buffered saline (PBS), and supplemented with a cation chelator (EDTA). The cells are not left in contact with the enzyme for more than the prescribed amount of time since these enzymes are capable of damaging cell membranes. Termination of enzyme activity is accomplished by neutralizing the activity of the solution by removing and adding medium containing fetal bovine serum. Figure 9 diagrams the protocol generally used for enzymatic dispersion of cell monolayers.

After counting, the cell suspension is diluted and the cells are inoculated (seeded) into appropriate culture vessels. A general rule is to seed 10^4 cells per 10 cm^2 area, a seeding density which corresponds from 1/4 to 1/3 the confluent density, thus allowing for optimum cell proliferation. The newly seeded culture vessels are incubated undisturbed for 24 hours to facilitate cell attachment.

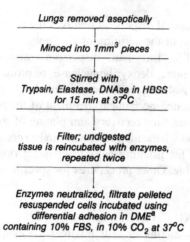

Figure 8.   Procedure for enzymatic isolation of cells from adult rabbit lung. [a]Indicates Dulbecco's modified Eagle's medium. (Adapted from S. Houser et al., ATLA, 17, 301, 1990.)

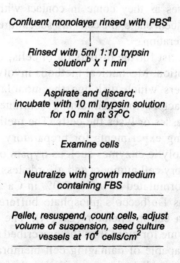

Figure 9.   Outline of the protocol generally used for enzymatic dispersion of cell monolayers. [a]Indicates PBS, Phosphate buffered saline without Ca^{+2} or Mg^2. [b]Indicates 1:10 dilution of 10× trypsin-EDTA concentrate.

D. MEASUREMENT OF GROWTH AND VIABILITY

There are important considerations when determining the integrity of a cell line or a newly established primary culture. The judgment required is generally based upon knowledge and experience that the individual has with cell culture in general. Overall, several criteria are assessed to determine the potential of the cells for future toxicological investigations, including the viability of the cultures; the appearance of the cells under the phase contrast microscope, especially as a finite cell line approaches the anticipated maximum number of doublings; and the ability of the cells to retain as many of the morphological and biochemical features as the original parent cell or primary culture. Most of these properties are discussed in subsequent chapters, when toxicological evaluations depend on the retention of these features.

More common criteria used to monitor cell growth and viability on a daily basis include cell counting (as described above, Figure 9); the exclusion of the vital dye, trypan blue; the ability to survive normal culture conditions; and the percentage of cells inoculated into the culture vessel that will attach to the culture surface. Each method is performed as part of the routine daily monitoring of the status of the cultures. Most of these features can be determined under the phase contrast microscope, assuming that the individual responsible for the cultures has a familiarity with the cells in question. Figure 10 outlines the method of vital dye exclusion, and Figure 11 illustrates a typical growth curve when all conditions for optimum cell growth have been met.

E. FREEZING CELLS

After addition of dimethyl sulfoxide or glycerin to the freezing medium, most cell types can be frozen indefinitely and later thawed for continued cultivation. This is a prerequisite to the use of finite cell lines in standardized testing where frozen stocks of the same passage of a cell type may be cultured independent of time and place. Depending on the extent of shared use and availability to other investigators, the feasibility of procuring an ultralow temperature freezer should be investigated. These units maintain temperatures just below the critical point of ice formation (-130°C) and are equipped with alarms and backup power units in the event of mechanical failure. For one or two individuals, a liquid nitrogen refrigerator is convenient but requires a standing order for the liquid due to constant evaporation.

Cell strains, cell lines, or primary cultures are cryopreserved when the optimal conditions for the culture have been attained. The procedure for preservation begins as described above for subculturing of monolayers (see Section VI.C, Subculturing). Cell monolayers are dispersed by trypsinization, neutralized with growth medium containing fetal bovine serum (complete growth medium), and centrifuged at $400 \times g$ for 10 min. The supernatant is discarded and the pellet is resuspended in complete growth medium containing 7.5% DMSO at a concentration of 1 to 3×10^6 cells per milliliter in sterile, 2ml, screw-capped cryoampules. The ampules are then labeled and frozen at a controlled rate of -1°/min in an automated freezing chamber. These chambers are usually available as accessory units accompanying liquid nitrogen refrigerators. Alternatively, the ampules may be frozen in the vapors of liquid nitrogen by placing them in a styrofoam container and positioning it in the neck of the liquid nitrogen refrigerator overnight. It is important that the level of the

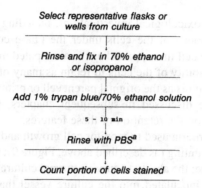

Figure 10. Outline of trypan blue vital dye exclusion method. [a]Indicates phosphate buffered saline.

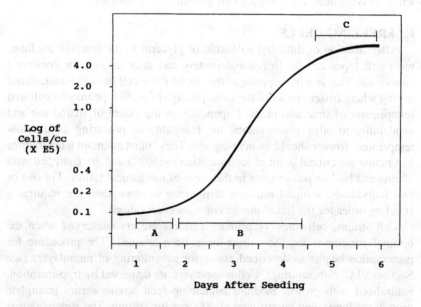

Figure 11. Diagram of a normal cell cycle during optimum culture conditions. (A) The *lag phase* immediately follows initial seeding of the tissue culture plates with cells and corresponds to the time period required for cells to settle, attach, and acclimate to the culture environment; (B) the *logarithmic phase* represents a rapid growth stage where the cells may advance through two to four population doublings; (C) the *plateau phase* is achieved when the monolayer is at confluency and the cells show contact inhibition. Proliferation essentially ceases at this stage.

nitrogen not be too high so as to freeze the cells too rapidly. The next day, the ampules are snapped into cryocanes and immersed in the liquid.

Adequate cell stocks of each frozen passage should be maintained and a representative vial of cells is thawed and cultured soon after freezing to evaluate the freezing process. This will ensure against the loss of a line, should contamination or accidents occur. An ampule is thawed rapidly by placing it in a shaking water bath, or manually rotating the vial, at 37°C for 3 to 4 min. The outside of the ampule is wiped with a 70% alcohol swab, allowed to dry, and the contents are pipetted into warm growth medium using a sterile plugged Pasteur pipette. For most cell types, the culture may be left undisturbed in the incubator in the appropriate atmosphere overnight, after which the medium is replaced.

F. CELL IDENTIFICATION

In a laboratory that is accustomed to handling several cell lines by various individuals, it is necessary to verify the original cell type in use. There is a high risk of cross contamination of cell lines in such situations. Some methods used for identification of characteristic cell markers include labeling with fluorescent antibodies, cytogenetic karyotypic studies and isoenzyme profiles. In addition, it is also necessary, on occasion, to check for the possibility of specific mycoplasma infections of the cultures. This contaminant is described as a bacteria, lacking a cell wall, which tenaciously adheres to cell membranes and cannot be eliminated with routine addition of antibiotics to the media. In fact, the indiscriminate addition of penicillin and streptomycin to the growth and saline-buffered wash media have been blamed for mycoplasma contamination. All of these tests are available commercially for a fee through the mail. Alternatively, the task of verifying a cell line or testing for mycoplasma can be assigned to a conscientious graduate student.

G. SOURCES OF CELL LINES

An alternate source of cell lines is a certified cell bank, such as the American Type Culture Collection (ATCC, Rockville, Maryland, U.S.). This facility operates as a non-profit institution and stocks thousands of cell types classified as either certified cell lines (CCL) or cell repository lines (CRL). CCLs have more information available as to the donor, karyotypic analysis, expected population doublings, cell structural features, and enzyme markers. In addition, the ATCC also houses repositories for viruses, fungi, bacteria, genetic probes, and vectors.

Also, the National Disease Research Interchange (NDRI, Philadelphia, PA) was established in 1980 as a non-profit, independent corporation. Its objectives are to procure, preserve, and distribute human cells, tissues, and organs for biomedical research.

REFERENCES

1. **Rinaldini, L. M.**, The isolation of living cells from animal tissues, *Int. Rev. Cytol.*, 7, 587, 1958.

2. **White, P. R. and Waymouth, C.**, Chemically defined synthetic nutrients for tissue cultures for pharmacological studies, *Ann. N.Y. Acad. Sci.*, 58, 1023, 1954.

3. **Bashor, M. M.**, Dispersion and disruption of tissues, in *Methods in Enzymology, Cell Culture*, Vol. 53, Jakoby, W. B. and Pastan, I. H., Eds., Academic Press, New York, 1979, chap. 9.

4. **Freshney, R. I.**, Introduction: Principles of sterile technique and cell propagation, in *Animal Cell Culture, A Practical Approach*, Freshney, R. I., Ed., IRL Press, Oxford, U.K., 1986, chap. 1.

5. **Barnes, D. W., Sirbasku, D. A., and Sato, G. H.**, Cell culture methods for molecular and cell biology methods for preparation of media, supplements, and substrata for serum-free animal cell culture. I, *Methods for Serum-Free Culture of Cells of the Endocrine System*, A. R. Liss, New York, 1984.

6. **Patterson, M. K.**, Measurement of growth and viability of cells in culture, in *Methods in Enzymology, Cell Culture*, Vol. 53, Jakoby, W. B. and Pastan, I. M., Eds., Academic Press, New York, 1979, chap. 11.

7. **Grand Island Biological Company (GIBCO)**, Catalog, Grand Island, NY, 1993, pp. 82 and 119.

8. **Ambesi-Impiombato, F. S., Parks, L. A. M., and Coon, H. G.**, Culture of hormone-dependent functional epithelial cells from rat thyroids, *Proc. Natl. Acad. Sci. U.S.A.*, 77, 3455, 1980.

9. **Masui, H., Miyazaki, K., and Sato, G. H.**, Culture of human lung carcinoma cells in serum-free medium, in *Hormonally Defined Media — A Tool in Cell Biology*, Fischer, G. and Wieser R. J., Eds., Springer-Verlag, New York, 1983, p. 430.

10. **Barnes, D. and Sato, G.**, Methods for growth of cultured cells in serum-free medium, *Anal. Biochem.*, 102, 255, 1980.

11. **Barnes, D. and Sato, G.**, Serum-free cell culture: a unifying approach, *Cell*, 22, 649, 1980.

12. **Orly, J., Weinberger-Ohana, P., and Farkash, Y.**, Studies on regulation of ovarian steroidogenesis in vitro: the need for a serum-free medium, in *Hormonally Defined Media*, Fischer, G. and Wieser R. J., Eds., Springer-Verlag, New York, 1983, p. 274.

13. **Gilchrest, B. A.**, Defined culture systems for pharmacological and toxicological studies, *Br. J. Dermatol.*, 115, 17–23, 1986.

14. **Barile, F. A., Ripley-Rouzier, C., Siddiqi, Z-e-A., and Bienkowski, R. S.**, Effects of prostaglandin E1 on collagen production and degradation in human fetal lung fibroblasts, *Arch. Biochem. Biophys.*, 265, 441, 1988.

15. **Houser, S., Borges, L., Guzowski, D. E., and Barile, F. A.**, Isolation and maintenance of continuous cultures of epithelial cells from chemically-injured adult rabbit lung, *ATLA*, 17, 301, 1990.

Chapter 2

MECHANISMS OF CYTOTOXICOLOGY

I. INTRODUCTION

It is useful to separate cell culture methods in toxicology into two different areas of application: 1) *Studies of mechanisms* of toxic effects, including effects on cell membranes, DNA, protein, and lipid production; mitochondrial activity, lysosomal function, and other secretory pathways; and biotransformation of chemicals; and 2) *Testing*, where the qualitative and quantitative toxicity of previously unclassified chemicals is estimated.

The mechanistic studies often follow conventional animal toxicological studies.[1-4] In this situation, cell cultures are used to analyze an effect which has already been studied pathologically or biochemically in the animals. For example, the liver toxicity of acetaminophen was detected clinically and confirmed using animal tests. Hepatocytes and liver parenchymal cells were then used to identify the mechanism of toxicity. In such studies, the choice of the culture system is determined by the specific question. The investigator has the flexibility to select the system necessary to answer the question as efficiently as possible.

When using cell culture methods for *testing*, however, it is important to continue to use exactly the same system, and one which can be compared to standardized procedures in other laboratories. In addition, the results of a selected system should be evaluated to determine the relevance of the data for a specific type of human toxicity. Thus, research aimed at elucidating the mechanisms of toxic effects allows for a wide variety of test systems that can be applied; whereas research for testing purposes requires a limited number of well defined, standardized cell culture systems. Nevertheless, these two areas of application in toxicology have complemented each other. For instance, established test systems have been developed as a consequence of mechanistic cell culture studies within specific fields.[5]

Certain disciplines in toxicology have utilized cell culture methods early during the development of the techniques, including studies of carcinogenicity, mutagenicity, fetal toxicity, and immunotoxicity. As the protocols and technology progressed, it was possible to culture pure populations of differentiated human or animal cells from most organs, even if viability was of short duration. The development of cellular testing systems, however, has been more sluggish and erratic. In the 1950s, cell cultures were used to evaluate local irritancy of dental and surgical materials. Ten years later, cellular assays were instrumental in identifying chromosomal aberrations and mutations. In the 1970s, cellular carcinogenicity assays were developed, and more recently, *in vitro* test methods for acute systemic toxicity, teratogenicity and immunotoxicity have surfaced. This section, therefore, will present a classification of the principles of

27

general systemic toxicity and organ specific toxicity and their underlying mechanisms, as they apply to cells in culture.

II. CLASSIFICATION OF CELLULAR TOXICITY

A. FUNCTIONAL CLASSIFICATION

Other than the ability of a chemical to possess genotoxic or carcinogenic potential, cellular toxicity may be classified in several ways. One example is the distinction between local and systemic toxicity. Other classifications include, but are not limited to: acute vs. chronic toxicity, immediate vs. delayed toxicity, diffuse vs. targeted toxicity, and high dosage vs. low dosage toxicity.[6-8]

Figure 1 depicts the classification of primary toxic mechanisms based on different *levels of organization* in the human body which could be affected by a chemical. The toxic substance must attack any of the following levels.

1. Basal Cell Functions

These fundamental processes involve structures and functions common to all cells in the human body, including the processes necessary to maintain cell membrane integrity, mitochondrial oxidation, translation and transcription, and lysosomal enzyme activity. In cultured cells, these functions are generally delegated to undifferentiated or dedifferentiated cell lines which have been maintained in continuous culture over several passages. Yet some cell lines do not require several passages but may dedifferentiate after several hours in culture. Hepatocytes, for instance, lose the high initial levels of enzyme activities after several hours in culture. These cell lines, however, are not necessarily without specific functional characteristics. Thus, most cell lines of mesenchymal origin (fibroblasts), for instance, maintain their protein secretory activities, although many of their original morphological and biochemical features have been lost. It must be remembered that these types of continuous cells have been traditionally used for mechanistic pharmacological and toxicological studies. Thus, regardless of their organ of origin, finite or immortal dedifferentiated continuous cell lines have historically been grouped into different toxicological and pharmacological categories because of their varied *mechanistic* response to chemicals. Cell lines with specialized persistent metabolic functions, such as MDCK kidney cells or neuroblastoma 1300 cells, demonstrate reactions to toxic agents based on specialized functions which have persisted to some extent. In cytotoxicity testing, however, most chemicals precipitate the same toxic alterations of basal cell functions in cell types of different origin, as well as in cells which are similar to those lacking most cell-specific functions.

2. Origin-Specific Cell Functions

These structures and functions distinguish cells of varied origin and label them as specialized cells with origin-specific cell functions. These include

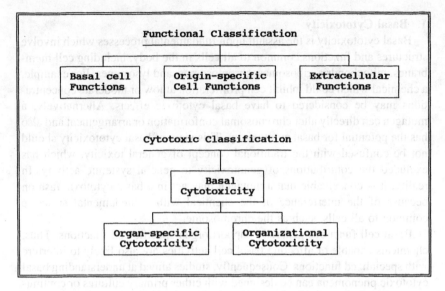

Figure 1. Areas of toxic mechanisms based on levels of organization.

structures which are unique to the cell, help in identifying the cell type, and may require an energy source in addition to their basal functions. An example of such structures and functions include: the presence of cilia (tracheal epithelial cells), contractility (myocardial cells), hormone production (cells of glandular or intestinal origin), and cells with high enzymatic activity capable of demonstrating chemical detoxification (hepatocytes or lung cells of epithelial origin).

3. Activities Destined for Extracellular Functions

Certain cell processes and functions produce mediators destined for extracellular activities. Their role in cytotoxic classification, therefore, is limited primarily because interference with their activity at the destination does not reflect their cell origin. A secretory process, such as the release of neurotransmitters by neuronal cells, or secretion of collagen by lung fibroblasts, may be difficult to trace at the target organ or cell within a cell culture system. The synthesis of extracellular matrix components is another example where interference by a toxic agent at the sight of deposition does not reflect the effect of that chemical on the epithelial cell. Of greater value as an indicator of toxicity may be the interference with these extracellular components by chemical substances either by nonspecific binding to the component or blocking receptors on the target cell. Nevertheless, the measurement of these functions *in vitro* may overlap with origin-specific functions, and thus may be an indication of organ-specific cytotoxicity (see below).

B. CYTOTOXIC CLASSIFICATION

Analogous to the functional classification, there are three types of mechanisms for toxicity, and thus a parallel system for cytotoxic classification can be described.[5-8]

1. Basal Cytotoxicity

Basal cytotoxicity is the assault upon fundamental processes which involve structures and functions common to all cells in the body, including cell membranes, mitochondria, ribosomes, chromosomes, and lysosomes. For example, a chemical which could inhibit cell proliferation at low or moderate concentrations may be considered to have basal cytotoxic effects. Alternatively, a mutagen can directly alter chromosomal conformation or arrangement and also has the potential for basal cytotoxicity. This view of basal cytotoxicity should not be confused with the traditional concept of general toxicity, which has excluded the contributions of genotoxicity to general systemic activity. In reality, it is conceivable that a mutagen may act in a basal cytotoxic fashion because of the interference of the chemical with a fundamental structure common to all cells, such as the chromosome.

Basal cell functions generally support organ-specific cell functions. Thus, chemicals capable of affecting basal cell activities are also likely to interfere with specialized functions. Consequently, studies aimed at understanding basal cytotoxic phenomena can be designed with either primary cultures or continuous cell lines and cell strains. Continuous cultures offer the advantages of working with more homogeneous populations, are well characterized, and are more easily cultivated when compared to primary cultures. Among continuous maintenance cultures, cells of embryonic or adult neoplastic origin are more easily manipulated and afford longer *in vitro* life spans. Although many continuous cell lines retain highly specialized biological activities *in vitro* that are characteristic of their original tissues, there is some loss of differentiation with continued passage.

Basal cytotoxicity from a chemical, therefore, can be demonstrated as a result of chemical attack upon cell membrane integrity, mitochondrial activity, or protein or DNA synthesis, since these are examples of fundamental metabolic functions which are common to all cells. In fact, the metabolic activity of cultured cells may reflect their adaptation to *in vitro* conditions, rather than their primary site of origin, which may account for different cell types displaying similar responses to toxins.[9] The level of activity in these cells, however, may vary from one cell type to another, depending on the cell's metabolic rate and contribution to homeostatic mechanisms, but the processes do not differ qualitatively among organs. That is, if chemical injury in an animal or human is a representation of the failure of the entire system to support the target organs, then all cells will display similar dysfunctions in their ability to maintain homeostasis. In many cases, target organ toxicity in humans may be caused by basal cytotoxicity of chemicals distributed to the corresponding organ. If this hypothesis is true, tests using primary cultures and continuous differentiated cell lines could be developed to cover a large percentage of toxic effects, and would reduce the need to introduce many laborious systems with organ-specific cells which are structurally and mechanistically unrelated.[10]

In order to show that a response to a toxic agent by a cell line is an expression of a basal cytotoxic phenomenon, several criteria must be met:

1.   A substantial number of chemicals of various molecular categories should be tested and shown to induce a degree of toxicity at moderate concentrations, especially those which are known to be toxic *in vivo*.
2.   Metabolic functions of various cell lines removed from the parent cells should show similar responses to toxic agents. This has been demonstrated in a variety of multicenter studies.[16,19,20]
3.   Cytotoxic agents with similar mechanisms, which show the same cytotoxic response *in vivo* and *in vitro*, would support the contention that this response is due to basal cytotoxicity.

Such experimental designs have also recently been incorporated into multicenter studies and form the basis of current validation programs (see Chapter 11).

Historically, there have been several objections to the use of cultured cells, in general, and basal cytotoxicity testing, in particular, for toxicity testing and screening. Among these, the most important objections are based on current technical limitations which have not been fully investigated, including:

1.   Some types of cells require hormonal, neuronal, and immunological adjustments, which are not readily present in routine monolayer cultures. Models designed to mimic extracellular functions and organizational toxicity may improve the ability to distinguish among the three levels of cytotoxicity.
2.   Many continuous cell lines are not capable of metabolizing xenobiotics. As a result, the effects of chemicals which require biotransformation for their toxicity will not be observed. The use of organ-specific primary cultures that retain high levels of enzymatic activity may overcome this drawback.
3.   Insufficient attention has been devoted to the significance of exposure time. Currently available tests measure cytotoxicity according to three types of exposure: (a) short-term tests involving exposure for periods up to 4 hours; (b) long-term tests, for periods of 24 or more hours; and (c) special tests involving unstable, volatile or insoluble materials.[6] The criticisms have implied that all these periods of exposure represent short incubation and observation times and may be predictive for acute *in vivo* effects only. In fact, cell culture systems may be modeled to mimic chronic exposure *in vivo* through manipulation of subcultivation parameters (see Section III.C below for a discussion of chronic exposure).

2. Organ-Specific Cytotoxicity

Chemicals which display selective toxicity, especially when *in vitro* concentrations are lower when compared to *in vivo* concentrations, qualify for organ-specific toxicity. Organ-specific cytotoxicity, therefore, is the effect of a chemical on a specialized function characteristic of a particular cell type, i.e., origin-specific cell function. Measurement of origin-specific cell functions, and its extrapolation to organ-specific toxicity, requires the use of primary cultures. Primary cell cultures have morphological and biochemical features that are similar to those of the original tissue. Therefore they offer the possibility for comparative studies of specialized tissues obtained from a variety of mammalian species. Primary cultures are generally more sensitive to the effects of toxic chemicals than cell lines because of their simultaneous role of adapting to the cell culture environment while being exposed to a toxic substance. The main limitations of primary cultures, however, include the difficulty of technically establishing the culture, the low cell yields, the lack of homogeneity, and the rapid loss of organ and cell-specific markers which determine organ-specific cytotoxicity.

Depending on the cell type, the use of primary cultures can give some assurance as to the maintenance of specialized functional markers present in the parent cell. Thus, by comparing similar classes of chemicals against primary and continuous cultures from the same organ, the differentiation of basal from organ-specific cytotoxicity is realized. In addition, primary and continuous cultures from the same organ can be incubated with structurally similar chemicals, and can also be used to distinguish a substance with a predilection for a particular organ. For example, in order to screen hepatotoxic agents, it is necessary to use a combination of primary hepatocyte cultures and parallel cultures of dedifferentiated cells. If a considerably lower concentration of a substance is needed to inhibit 50% of proteins synthesized (IC_{50}) in the primary hepatocyte culture than in an analogous parenchymal hepatoma cell line, this would suggest that the chemical is hepatotoxic. Similar moderate concentrations of a particular chemical causing a toxic response in both types of cultures could be explained as a basal cytotoxic phenomenon.

3. Organizational Toxicity

Chemicals which interfere with products of metabolism or secretion are considered to interfere with organ-specific functions as well as organizational functions. Tubocurare, for instance, causes respiratory paralysis *in vivo* by interfering with the action of the chemical neurotransmitter, acetylcholine, at the neuromuscular junction. Tubocurare, therefore, satisfies the requirements for classification as a chemical that causes organizational toxicity. Specific neurotoxicity *in vivo* is mostly a result of inhibition or interference with neurohumoral transmitters. In addition, endothelial cells selectively screen and prevent large molecules from entering the spinal fluid, thus forming the blood-brain barrier. Therefore, *in vitro* tests to screen for this type of organizational toxicity should mimic barriers capable of preventing the transfer of large molecules across membranes. Models, such as co-cultures on either side of Millipore® filters, have been used to predict the toxicity of chemical metabolites or secretory products of cells and their effect on the monolayer lying below the filter. Models for intestinal absorption have also employed this system. In addition, agents capable of preventing the binding of macro-molecules with specific cell receptors qualify as interfering with organizational toxicity.

These classifications are important for understanding the possibilities of cellular methods in toxicology. Basal cytotoxic mechanisms can be studied in systems with undifferentiated finite or continuous cell lines, while organ-specific cytotoxicity must be studied in primary cultures with well-differentiated cells from different organs. Organizational toxicity may be studied indirectly in cell cultures by examining the substrates or products of cell metabolism. In the example above, a chemical may be determined to interfere with the ability of a neuron to secrete acetylcholine by measuring the release of the neurotrans-mitter. It is important to note, however, that the basal functions in the cell are a prerequisite for the specific functions. That is, basal cytotoxicity may *imitate* the effects of organ-specific cellular toxicity or even organizational toxicity, by indirectly affecting the specific functions. Also, the mode of action of a chemical can be specific *within* each type of toxicity. For example, basal cytotoxicity can be precipitated by a lipophilic chemical either by a specific action on the sodium ion pump, or as a result of a nonspecific dissolution of the chemical in the nonpolar components of the cell membrane. The net result of either mechanism is loss of membrane integrity.

III. MECHANISMS OF GENERAL TOXICITY

A. LD$_{50}$ TEST IN ANIMALS

Protocols designed to measure systemic toxicity in animals are critical to the study of toxicology. This axiom has gained fundamental acceptance, in part because animal testing is usually the first step in a series of programmed experiments when testing is required. In addition, systemic toxicity is also useful as a preliminary screen so that chemicals may be classified according to their risk potential. The toxic potential of industrial chemicals and household products is often assessed this way by using standard animal models for systemic toxicity. These models comprise the basic tests used for risk assess-ment; that is, attempting to predict the toxic potential of a chemical to humans by extrapolating the results obtained from the animal experiments.

The LD$_{50}$ test is an example of a method frequently used for its ability to screen substances for their toxic potential. The method is designed to determine the dose of a substance which kills 50% of the animals to whom it is admin-istered. The protocol may be supplemented with simultaneous physiological observations and histopathological studies. One experiment can be very costly, however, consisting of 10 to 20 groups of rodents with 20 animals per group. This design accommodates testing for increasing doses of the chemical to be used in each group and ensures that the desired dose, which proves to be effective in 50% of the subjects, is bracketed. The procedure may also be painful to the animals, especially at sublethal doses. Figure 2 outlines a typical protocol for the determination of the LD_{50} of a substance.

Currently, the popularity of the test is not based on its scientific validity but more on its acceptance as a standard requirement of regulatory agencies. Nevertheless, many laboratories have instituted programs to replace acute toxicity tests on animals with cell culture methods. Until recently, this endeavor has encountered considerable resistance for several reasons. First, common knowledge dictated for many years that a large proportion of chemicals exhibited their toxicity through organizational mechanisms (see Section II above). With this bias, cell culture mechanisms are not seen as valuable because of the lack of understanding of the toxic events which occur with isolated cells. Also, it was believed that most cytotoxic chemicals could only be tested using expensive combinations of cells in primary culture. In addition, biotransformation has always played an important role in explaining general toxicity. Cell culture systems, therefore, were not considered as methods which could contribute to the pharmacokinetic data. The explosion of knowledge gathered from cellular techniques has negated most of the unsupported biases.

B. ASSUMPTIONS CONCERNING THE USE OF CELL LINES

By using cellular and subcellular model systems, perfused organs, and other conventional isolated tissue techniques to supplement animal experimentation, it is possible to gain a deeper knowledge of primary toxic mechanisms. The earlier body of research describing toxicity by physiological and pathological methods is continuously being replaced by the methods of cell biology and biochemistry. In addition, pharmacokinetic data supplements the information obtained from the cellular effects of a chemical and contributes to the knowledge of the mechanism of human toxicity.

Cellular functions present in the original isolated parent cells do not guarantee that organ-specific effects of chemicals will be observed in the subsequent generations with passage. Flint[21,22] has argued that basal cytotoxicity tests, especially those involving permanent cell lines, are of limited value, because the cells have metabolically distanced themselves from their tissues of origin. Established cell lines, whether continuous or finite, differentiate and lose their organ-specific functions. Conversely, the systems are actually designed to measure basal, and not target organ, cytotoxicity; that is, with the loss of specific functions, the toxin in question affects metabolic processes by altering mechanisms important to maintain basal activity. As a consequence, in order to assess organ-specific cytotoxicity, primary cultures must be used.[11-12] One way of distinguishing basal cytotoxicity from target organ selectivity of a chemical is by comparing the data obtained with cell lines and primary cultures from the same organ.

Cell lines can also be used for mechanism studies for chemicals which are known to affect basal cell functions. Such substances include: antibiotics, chemotherapeutic agents, digitalis, organic solvents, and heavy metals. This is

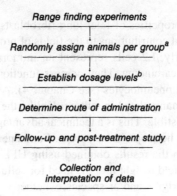

Range finding experiments

Randomly assign animals per group[a]

Establish dosage levels[b]

Determine route of administration

Follow-up and post-treatment study

Collection and
interpretation of data

Figure 2. Outline of LD_{50} protocol. [a] In the case of rodents, assignment is based on a minimum of 10 rodents per group. [b] Usually 5 to 10 dosage levels with ratios of 1.5 to 2.0 between successive doses.

an advantage of such a system because methods incorporating cell lines are simpler, do not require repetitive sacrificing of animals, and use cells which are similar from one experiment to the next. Conversely, the primary cultures vary in several aspects, partly because the explants are taken from different human or animal donors principally because of the different ages or group of animals, respectively. These problems can be minimized by consistently using the same species and age of animals. Freshly isolated cells may adapt differently to the conditions in the primary culture at different times of isolation, but this inconsistency may also be reduced by rigorously standardizing the techniques.

The use of established cell lines has traditionally been credited with identifying the mechanisms of cellular toxicity. The inference has been added that the cell line is also a reflection of the organ of derivation. This concept, however, is subject to question, especially since it is generally accepted that, with time and passage, primary cultures lose their original characteristics inherited from the parent cell *in vivo*. Thus, it is appropriate to use continuous cell lines for testing general toxicity, rather than assuming that the markers they retain in culture are a reflection of the organ of origin.

C. USE OF CELL LINES FOR ASSESSING GENERAL TOXICITY
1. Continuous Cell Lines

Continuous cell strains and cell lines have traditionally been used for studies of cell biology, biochemistry, and more recently, toxicology. An example of the type of cells that have broached the different disciplines includes human fetal lung fibroblasts. These cells, such as HFL1, WI38 and MRC-5 strains, have been extensively characterized and have been used for identifying the mechanisms underlying the cell biology of human lung and connective tissue disease.[13] They have a diploid phenotype and are described as having a finite *in vitro* life span, yet can be maintained in culture for many population doubling levels. The properties of fetal lung fibroblasts have allowed for the realization of important contributions to the fields of protein biochemistry and cellular aging. Similarly, L2 cells represent an immortal epithelial cell line derived from rat lung carcinoma. They possess functions and features which mimic alveolar type II pneumocytes (see Chapter 4). As with the fibroblasts, these cells are easily manipulated in culture. Unlike the fibroblasts, however, their supply is inexhaustible. This is a technical advantage although it must be noted that they have a heteroploid karyotype. Fundamentally, the similarities or differences between the results obtained using HFL1 and L2 cells to test toxic chemicals may yield insight into the need for cellular systems to include cell lines of various origins and/or species, whether or not they are derived from the same organ. If *in vitro* cell growth, for example, is truly a universal phenomenon common to all cells (assuming they demonstrate some predictable pattern of mitotic division), then the response of the cell lines to a chemical should be similar. This would lend support to the hypothesis that continuous cell lines are more inclined to detect basal cytotoxicity only, independent of the organ of origin.

2. Acute vs. Chronic Cell Tests

A major criticism of *in vitro* cytotoxicity assays is that the procedures measure only acute toxicity, primarily because the exposure is of short duration and occurs through one cellular passage level. This is because occasionally it is difficult to determine the difference between acute and chronic exposure in culture. Since one culture cycle (one passage level) can extend from three to seven days depending on the cell line, it is conceivable to extend the exposure time, within the time of the cycle, to determine the response to a chemical. In this way, exposure to the chemical through at least one cell cycle may differentiate between acute and subacute, or subacute and chronic. This sequence for exposure would also reveal any tolerance developed against a chemical.

The effects of chemicals *in vivo* at the cellular level may also result from repeated exposure and may follow familiar patterns, including: a) the slow accumulation of the chemical in target tissues until acute toxic concentrations are reached; b) the slow accumulation of the chemical in blood which subsequently distributes to target tissues; and, c) the chemical does not accumulate but causes toxicity through repeated insults until threshold injury results. To address these mechanisms, repeated and chronic exposure assays could be performed, but must be carefully outlined so as to mimic the *in vivo* situation. Examples of such experiments may be designed as follows:

1. To determine if a chemical can elicit a toxic response by slow accumulation in the target tissue (pattern "a" above), cultures are exposed to *low but increasing doses* of the chemical for at least two to three passages. In this experiment, it can be determined if the chemical causes toxicity at concentrations below threshold or if it is necessary to accumulate the concentrations prior to the toxic response.

2. In prolonged repeated intermittent exposure experiments, designed to mimic pattern "b", proliferating cells are continuously exposed to *low* doses of the chemical for three or more passages. The mechanism of accumulation at low concentrations and the corresponding toxic response is then measured. Its effect on mitosis can also be observed.

3. Repeated dosage assays are designed to address mechanism "c". These experiments involve short exposure to the test chemical (1 to 24 hours), followed by substitution with fresh medium, thus allowing sufficient time for the uncoupling of a non-covalently bound substance. Incubations are repeated in each of 3 successive passages (each passage is about 2 to 3 doubling levels). The objective is to determine the extent of insult occurring at the cellular level in the absence of the agent, after repeated insult.

D. APPLICATIONS OF BASAL CYTOTOXICITY DATA

1. In Human Risk Assessment

Basal cytotoxicity data (BC-data) has the potential to be a powerful tool for human risk assessment. BC-data represents *in vitro* toxicity of chemicals to cultures of non-differentiated cells, such as finite or transformed cell lines. Animal cell lines from various origins may also be used but the current understanding is that cells from human origins may reflect greater sensitivity to cytotoxicity and may facilitate *in vivo* comparisons. As outlined above, BC-data encompasses various toxic parameters, such as the determination of inhibitory concentrations at several exposure levels with a variety of toxic criteria. These criteria include, but are not limited to, determination of protein content of cultures, neutral red uptake and MTT assays, and LDH release. At present, BC-data is not considered part of the regulatory agenda needed to screen toxic substances before acceptance by industrial, scientific, agricultural and manufacturing communities. Results from current initiatives and validation efforts, however, indicate that BC-data may be useful in the future for human risk assessment.

Many studies in the last 10 years have indicated the usefulness of *in vitro* BC-data for the prediction of human toxicity. In general, these studies compared toxicity using cell lines with concentrations derived from human and animal exposures.[5,7,15] These exposures involved acute systemic and target organ toxicity, eye and skin irritancy, and animal teratogenicity. The BC-data most often analyzed includes 24- and 72-hour 50% inhibitory concentrations (IC_{50}s) for animal or human cells. Some of the salient features of the studies are summarized:

1. Good correlations have been computed in several studies between IC_{50}s determined *in vitro* and rodent LD_{50}s for a total of several hundred chemicals. Hopkinson et al.[10] have shown that IC_{50} values determined *in vitro*, using a continuous rat lung epithelial cell line, were as accurate in predicting human lethal concentrations of 50 chemicals as human equivalent toxic blood concentrations (HETCs) derived from rodent LD_{50}s.

That is, *in vitro* cytotoxicity testing procedures have the potential to predict human chemical toxicity at least as accurately as rodent LD_{50} determinations (Figures 3 and 4).

2.   Similarly, good correlations were determined between IC_{50} values and eye irritancy protocols.

3.   There were good correlations between various systemic target organ toxicities, such as toxicity to the kidneys, liver, nervous system, and IC_{50}s determined in continuous cell lines, indicating that most cases of target organ toxicity is reflected as basal cytotoxicity distributed to the organs.[6-10]

4.   There was a comparison of *in vitro* data and acute human lethal blood concentrations for a limited number of compounds.

5.   Several recent studies have indicated that IC_{50}s, derived from 24- and 48-hour studies using cell lines are similar, regardless of the origin of the cultures and the toxicity criteria used (it should be noted however that recent evidence suggests that human cell lines may be more suitable for detecting cytotoxicity than cells of animal origin).[11-16]

6.   Cultured cells are less likely to present false positive results, as long as the true IC_{50} concentrations are measured and the culture techniques are standardized, at least within a particular method.[16]

The experimental evidence supports the conclusion that regardless of the level of differentiation, the majority of cell lines respond to toxic concentrations of chemicals in a similar manner. This suggests that continuous cell lines can be used to assess general toxicity of the majority of substances.

Practically, cell line methods are viable alternatives to animal protocols and can be used for testing acute, systemic, and local toxicity. Along with primary cultures of organ-specific cells, the cell lines have traditionally been valuable in determining the mechanisms of action. Continuous cultures, however, may be more economical to use than the sacrificing of animals for primary cultures. Thus, this information appears to represent a maximal, potential expression of toxicity for most chemicals. In combination with toxicokinetic data, BC-data can predict most systemic target organ effects (such as hepatocellular, lung, and skin irritancy) and can be modeled for most general toxicity. BC-data cannot account for organizational or organ-specific effects, both of which are infrequently encountered in acute human toxicity of randomly selected chemicals. Moreover, BC-data may be supplemented by *in vitro* tests on isolated organelles or receptors, or organ-specific primary cultures, so as to aid in the screening of chemicals which do not show toxicity (that is, screen for false negative results). In fact, an analysis of *differential toxicity*, between BC-data and information derived from organ-specific or organizational databases, can be used to distinguish among the levels of cytotoxic classifications.

BC-data is more suited for risk assessment than most of the toxicity data generated in conventional protocols. For instance, by using human cells, species differences are essentially overcome and the information gathered can be

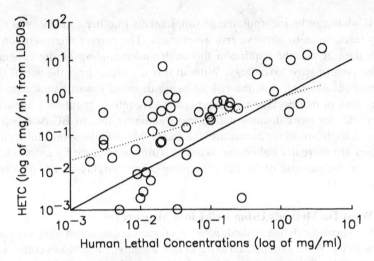

Figure 3. Graph of human equivalent toxic concentrations derived from rodent LD_{50} values vs. known human lethal concentrations for 50 chemicals. The regression line (dotted) has a slope of 4.4 ($r = 0.84$). The solid line represents a theoretical slope of 1.0. (From Hopkinson et al., *ATLA* 21, 167, 1993. With permission.)

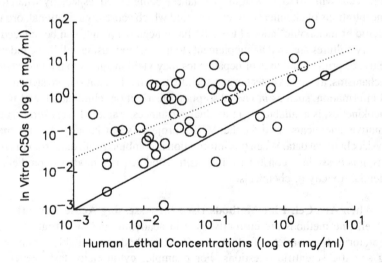

Figure 4. Graph of experimental *in vitro* IC_{50}s vs. known human lethal concentrations for 50 chemicals. The regression line (dotted) has a slope of 8.9 ($r = 0.48$). The solid line represents a theoretical slope of 1.0. (From Hopkinson et al., *ATLA*, 21, 167, 1993. With permission.)

compared directly to known human toxic blood and tissue concentrations. The latter analysis is possible especially when the compound, *in vivo*, is well absorbed, is not extensively metabolized from the parent molecule, and is not sequestered to a large extent in physiological compartments other than plasma.

BC-data can be incorporated as supplements to other conventional toxicity tests, so as to improve risk assessment. The correct interpretation of such data at present requires a thorough understanding of the concepts underlying *in vitro* toxicology. Without this understanding, the use of this type of information runs the risk of being directed towards unacceptable variations in interpretation among experts. Therefore, regulatory agencies must wait for more documentation and standardization of BC-data before general applications are prematurely initiated. Meanwhile, some multicenter studies are currently collecting, standardizing, and validating cytotoxicity data for the purpose of further presenting some insights on BC-data (see Chapter 11).

2. What Do Methods Using Cell Lines Measure?

Analytically, cell line methods measure concentrations which represent basal cytotoxicity. When extrapolated to the *in vivo* situation, these values correspond to human blood concentrations and may be interpreted as doses which pose a risk to all human organs. Accordingly, false positive results are not common (specificity of the method), unless the material is transformed *in vivo* to an innocuous metabolite. False negative findings (measuring sensitivity of the assay), however, occur with 10 to 30% of the substances being tested, especially when low concentrations of chemicals are encountered which cause organizational, organ-specific or metabolite-induced toxicity. False negative results can be reduced if primary cultures are used as supplements to the methods using cell lines. Indeed, the use of primary cultures of hepatocytes may yield insight into the underlying mechanisms, as well as detect organ-specific toxicity. In addition, physicochemical information, such as *in vivo* absorption, distribution, elimination, and pharmacokinetics, is available for individual substances. Table 1 lists several representative chemicals used in the MEIC project (see Chapter 11) and their physicochemical data. When this information is combined with data from *in vitro* tests, it is possible to construct very sensitive and accurate models to predict the potential toxicity of chemicals.

3. When Are Cell Line Methods Used for Assessing Acute Toxicity?

Cell line methods are currently used in academic and industrial research laboratories for a variety of situations. These methods are suitable for answering specific scientific questions. For example, cytotoxicity tests generally precede *in vitro* genotoxicity tests so that mutations and chromosome damage may be ascertained. Cytotoxicity testing is also becoming more popular with the development of new therapeutic drugs. Cell methods, including the use of primary cultures of hepatocytes and *in vitro* pharmacokinetic studies, are used as screening tests to assess acute and chronic toxicity. Setting the protocols in this sequence reduces the number of animals needed for *in vivo* testing. In addition, highly cytotoxic products are screened using cellular methods, thus eliminating the need to expose animals to these compounds.

TABLE 1
First Ten Chemicals Used for Cytotoxicity Testing
in the MEIC[a] Project and Corresponding
Physicochemical Data

Number	Chemical[b]	Mw	Mp	Log P	(Log P)^2	Log LC
1.	Paracetamol	151.2	151	0.80	0.64	−2.00
2.	Aspirin	180.2	135	1.22	1.49	−2.44
3.	Ferrous sulfate	151.9	–	–	–	−3.47
4.	Diazepam	284.8	125	2.80	7.84	−3.60
5.	Amitriptyline	277.4	196	4.92	24.20	−4.27
6.	Digoxin	781.1	265	1.26	1.59	−7.52
7.	Ethylene glycol	62.1	−12	−1.93	3.72	−1.32
8.	Methanol	32.1	−94	−0.71	0.50	−1.22
9.	Ethanol	46.1	−117	−0.31	0.10	−0.82
10.	Isopropanol	60.1	−89	0.05	0.00	−1.17

[a] Multicenter Evaluation for *In Vitro* Cytotoxicity.
[b] Mw, molecular weight; mp, melting point; log P, partition coefficient (log P octanol/water); LC, human toxicity data, acute lethal blood (molar) concentrations from clinical case reports.

From Hellberg, S., Eriksson, L., Jonsson, J., et al., *ATLA* 18, 103, 1990.

Similarly, there has been substantial effort into evaluating the potential predictive ability of general cytotoxicity tests for eye and local irritancy. This endeavor is already proving worthwhile for assessing the safety of cosmetic formulations, which would be tested in animals before human volunteer studies were undertaken.[23] Although such *in vitro* studies have not gained acceptance by the pharmaceutical industry, they can be used for human risk assessment as screening methods.

Cytotoxicity tests are also employed by chemical companies for the screening of products already marketed. These include pesticides, food and beverage additives and preservatives, liquids or gases determined to increase the risk of occupational hazards, and chemicals responsible for water, air and soil pollution. When these chemicals are introduced into the cellular systems, the concentrations necessary to produce toxicity can be compared to compounds of similar or identical classes. Consequently, toxicities are compared and the results are used to determine whether further testing is needed.

IV. MECHANISMS OF ORGAN-SPECIFIC TOXICITY

A. USE OF PRIMARY CULTURES FOR ORGAN-SPECIFIC TOXICITY

In traditional animal experimentation, the target organ for toxicity is identified and the cells are isolated and studied to determine the mechanism. Primary cultures of highly differentiated cells with specialized functions are used for this purpose. The effects of the toxic substance on specific functions are studied, while other manipulations of the methods, such as incubation times and choice of media, depend on the desired information. This situation may result in an inordinate number of methods included in the study. For mechanistic studies, human or rodent cells are used most often. Blood cells and surface epithelial cells are most accessible but are technically restricted to the study of acute toxic mechanisms because of the rapid dedifferentiation of the cells in primary culture. Certain cultures, however, can be kept in a functional state for months or years. For instance, reaggregated organ-specific cultures of nerve cells permit studies of chemically-induced chronic degeneration of myelin sheaths.

Several systems have been used for mechanism studies of acute toxicity. For instance, organ cultures of tracheal cross sections are useful for identifying effects on cilia by chemicals found in tobacco smoke, and phagocytic properties of lung macrophages are suitable for the demonstration of asbestos and mineral dust toxicity. Similarly, primary cultures of rhythmically contracting heart myocytes are ideal for examining cardiotoxic chemicals. Prolactin-producing pituitary cells are secretory cells which can be used to study the chemical inhibition of hormonal endocrine secretions. These and other organ-specific cell methods are described in Chapter 3.

B. SPECIALIZED FUNCTIONS OF PRIMARY CULTURES

Other specialized functions of primary cultures may aid in the understanding of organ-specific toxicity. For instance, one of the main functions of the type II alveolar pneumocyte in the lung is the synthesis, storage, and secretion of pulmonary surfactant. This material, which is unique to the type II cell and selected fetal lung fibroblasts, is composed of saturated and unsaturated phospholipids which act to decrease the cell-air interface in the alveoli, thus allowing for gases to permeate the thin alveolar membranes.[17,18] The effect of chemicals on phospholipid synthesis in primary cultures of epithelial cells may be more pronounced than their effect on the more universal mechanism of protein production. Hence, as an indicator of cell toxicity, lipid production may be useful to differentiate the lung-specific toxicity of pulmonary-specific chemicals from general basal cytotoxicity. Therefore, measuring a product of cell metabolism or catabolism in primary culture, such as phospholipids, contributes to our understanding of organ-specific toxicity. By comparing inhibitory concentrations (IC_{50} values) based on the percent of response for lipid synthesis against values for cell growth, the sensitivity of the methods for lung toxicity is ascertained. For example, if a significantly lower concentration of a particular chemical suppresses lipid production to a greater extent than cell growth, this may support the hypothesis that this chemical has a toxic predilection for lung cells. Therefore, a specialized function in primary culture which is more inhibited by a lower dose of a chemical than by a higher concentration which prevents cell growth, would indicate that the chemical has the potential to interfere with that particular organ-specific function.

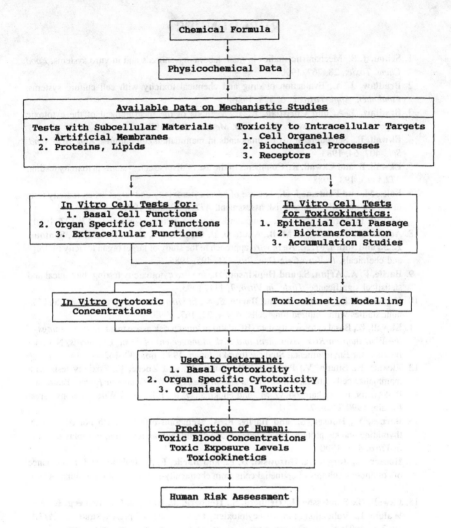

Figure 5. Organization and Applications of *In Vitro* Cytotoxicity Testing.

V. SUMMARY

Figure 5 summarizes the organization and applications of *in vitro* cytotoxicity testing and assigns a position for basal cytotoxicity, organ-specific, and organizational cytotoxicities within the general scheme. Overall, the objective of this information is to arrive at equivalent human toxic blood concentrations to assist in the prediction of human toxicity and evaluation of risk assessment. Note also that *in vitro* toxicokinetic studies contribute to the establishment of such a model. Together with *in vitro* concentrations derived from the tests, a model for cytotoxicity is formulated.

REFERENCES

1. **Schou, J. S.**, Mechanistic studies in man: laboratory animals and in vitro systems, *Food Chem. Toxic.*, 28, 767, 1990.
2. **Bradlaw, J. A.**, Evaluation of drug and chemical toxicity with cell culture systems, *Fundamen. Appl. Toxicol.*, 6, 598, 1986.
3. **Bradlaw, J. A. and Christian, R. T.**, Methods of the measurement of the cytotoxic response: a short review, *J. Tissue Cult. Methods*, 9, 1, 1985.
4. **Ekwall, B.**, Screening of toxic compounds in mammalian cell cultures, *Ann. N.Y. Acad. Sci.*, 407, 64, 1983.
5. **Ekwall, B. and Ekwall, K.**, Comments of the use of diverse cell systems in toxicity testing, *ATLA*, 15, 193, 1988.
6. **Balls, M. and Fentem, J. H.**, The use of basal cytotoxicity and target organ toxicity tests in hazard identification and risk assessment, *ATLA*, 20, 368, 1992.
7. **Ekwall, B.**, Screening of toxic compounds in tissue culture, *Toxicology*, 17, 127, 1980.
8. **Acosta, D., Sorensen, E. M. B., Anuforo, D. C., Mitchell, D. B., Ramos, K., Santone, K. S., and Smith, M. A.**, An *in vitro* approach to the study of target organ toxicity of drugs and chemicals, *In Vitro Cell. Dev. Biol.*, 21, 495, 1985.
9. **Barile, F. A., Arjun, S., and Hopkinson, D.**, *In vitro* cytotoxicity testing: biological and statistical significance, *Toxic. In Vitro*, 7, 111, 1993.
10. **Hopkinson, D., Bourne, R., and Barile, F. A.**, *In vitro* cytotoxicity testing: 24- and 72-hour studies with cultured lung cells, *ATLA*, 21, 167, 1993.
11. **Ekwall, B.**, Basal cytotoxicity data (BC-data) in human risk assessment, in *Proceedings of the Workshop on Risk Assessment and Risk Management of Toxic Chemicals*, National Institute for Environmental Studies, Ibaraki, Japan, 1992, pp. 137–142.
12. **Ekwall, B., Silano, V., Paganuzzi-Stammati, A., and Zucco, F.**, Toxicity tests with mammalian cell cultures, in *Short-Term Toxicity Tests for Non-Genotoxic Effects*, Bourdeau, P., Somers, E., Richardson, G. M., and Hickman, J. R., Eds., John Wiley & Sons, Great Britain, 1990, chap. 7.
13. **Barnes, Y., Houser, S., and Barile, F. A.**, Temporal effects of ethanol on growth, thymidine uptake, protein, and collagen production in human fetal lung fibroblasts, *Toxic. In Vitro*, 4, 1, 1990.
14. **Houser, S., Borges, L., Guzowski, D. E., and Barile, F. A.**, Isolation and maintenance of continuous cultures of epithelial cells from chemically-injured adult rabbit lung, *ATLA*, 17, 301, 1990.
15. **Ekwall, B., Bondesson, I., Hellberg, S., Hogberg, J., Romert, L., Stenberg, K., and Walum, E.**, Validation of *in vitro* cytotoxicity tests — past and present strategies, *ATLA*, 19, 226, 1991.
16. **Ekwall, B., Barile, F. A., Clothier, R. H., Gomez-Lechon, M. J., Hellberg, S., Nordin, M., Stadtlander, K., Triglia D., Tyson, C. A., and Walum, E.**, Prediction of acute human lethal dosage and blood concentrations of the first 10 MEIC chemicals by altogether 25 cellular assays, *In Vitro Cell. Dev. Biol.*, 26, 26A, 1990.
17. **Dobbs, L. G., Geppert, E. F., Williams, M. C., Greenleaf, R. D., and Mason, R. J.**, Metabolic properties and ultrastructure of alveolar type II cells isolated from elastase, *Biochim. Biophys. Acta*, 618, 510, 1980.
18. **Dobbs, L. G., Mason, R. J., Williams, M. C., Benson, B. J., and Sueishi, K.**, Secretion of surfactant by primary cultures of alveolar type II cells isolated from rats, *Biochim. Biophys. Acta*, 713, 118, 1982.
19. **Ekwall, B., Bondesson, I., Castell, J. V., Gomez-Lechon, M. J., Hellberg, S., Hogberg, J., Jover, R., Ponsoda, X., Romert, L., Stenberg, K., and Walum, E.**, Cytotoxicity evaluation for the first ten MEIC chemcials: Acute lethal toxicity in man predicted by cytotoxicity in five cellular assays and by oral LD_{50} in rodents, *ATLA*, 17, 83, 1989.

20. **Hellberg, S., Eriksson, L., Jonsson, J., Lindgren, F., Sjostrom, M., Wold, S., Ekwall, B., Gomez-Lechon, M. J., Clothier, R. H., Triglia, D., Barile, F. A., Nordin, M., Tyson, C. A., Dierickx, P., Shrivastava, R., Tingsleff-Skaanild, M., Garza-Ocanas, L., and Fiskesjo, G.**, Analogy models for prediction of human toxicity, *ATLA*, 18, 103, 1990.
21. **Flint, O. P.**, *In vitro* toxicity testing: purpose, validation, and strategy, *ATLA*, 18, 11, 1990.
22. **Flint, O. P.**, *In vitro* test validation: a house built on sand, *ATLA*, 20, 196, 1992.
23. **Balls, M., Reader, S., Atkinson, K., Tarrant, J., and Clothier, R.**, Non-animal alternative toxicity tests for detergents: genuine replacements or mere prescreens?, *J. Chem. Technol. Biotechnol.*, 50, 423, 1991.

Chapter 3

CELLULAR METHODS OF GENERAL TOXICITY

I. USE OF DIFFERENTIATED CELLS

A. INTRODUCTION

The data obtained from using animal experiments yield information pertaining to the dose for lethal or sublethal toxicity, which corresponds to many different general toxic mechanisms and effects. Similarly, cell systems also detect a wide spectrum of unspecified mechanisms and effects.[1] In contrast to animal experiments on general toxicity, however, all cell tests of general toxicity in current use measure the concentration of a substance which damages components, structures, or biochemical pathways within the cells. This range of injury is further specified by the length of exposure (the incubation time) to the chemical, thus allowing the test to be predictive for risks associated with toxic effects *in vivo*, for those doses of the tested substance. This assumes that similar concentrations can be measured in the corresponding human tissue for comparable exposure periods. It should also be possible to achieve estimated cytotoxic concentrations, in any human tissue, by local application or through parenteral administration. Consequently, the cell tests may be devised for assessing the potential for local irritancy. If undifferentiated cell lines are used in the test, the results will be limited to determining the risk for basal cytotoxicity of the measured concentrations. Conversely, the use of primary cultures of differentiated cells (such as those of liver, lung, heart, or kidney) will yield information capable of predicting cellular toxicity for each organ.

B. COMPARISONS TO MUTAGENICITY AND CARCINOGENICITY TESTING

Cell tests of general toxicity are different from cell tests of mutagenicity and carcinogenicity in many respects. The latter have the ability to suggest a known underlying mechanism, such as mutagenicity, transformation, promotion, induction, or progression, which is responsible for the chemical's toxicity. They are designated, therefore, as short-term tests and are applied to the prediction of cancer induction of chemicals in man. In comparison, assessing the potential carcinogenicity or mutagenicity of a chemical in humans, retrospectively, normally takes much longer. Cytotoxic tests instead simulate injury upon human tissues and organs from the tested substances, which may be caused by a number of incomplete mechanisms, during periods of exposure which are realistic for acute toxicity. This allows for direct extrapolation of the data from the quantitative test to the analogous *in vivo* situation. The cytotoxic tests are thus not short-term tests, but have the advantage of being inexpensive and less time-consuming in comparison to animal tests. There are also similarities between the short-term mutagenicity and cytotoxicity tests in cell cultures. The methods must be performed using a few standardized techniques whose relevance to human toxicity has been carefully evaluated. For mutagenicity tests, the relevance applies to the mechanisms that may be encountered in humans. For cytotoxicity tests, the concentrations must ultimately correspond to relevant human toxic or lethal blood concentrations.

Other differences between studies of toxic mechanisms in cell cultures and cytotoxicity testing exist. In the mechanism studies the measured parameters and systems vary, depending on the desired information. The empirical and statistical relevance of the system is secondary to understanding the mechanism of toxicity. The cell tests, however, must be standardized, be evaluated, and have well-defined criteria for toxic assessment, such as standards for cell viability and cell function. In addition, the protocols must be strictly followed so that interlaboratory variations are minimal.

II. METHODS USING CELL LINES

A. GENERAL TOXICITY CRITERIA

About twenty methods used for general toxicity testing have filtered through from those traditionally used in mechanism studies. Their validity has been established usually as a result of frequent use in one laboratory, followed by repetition and confirmation by others. This method of evaluation thus proceeds more as a matter of convenience for the laboratory, particularly if the assay can be easily accommodated to the routine of the testing facilities. The methods are necessarily similar — cells are exposed to different concentrations of the substance for a predetermined period of time, after which the degree of inhibition of viability or functional condition is measured. This criteria for standard of measurement then becomes the toxic end-point. The most common criteria for cell lines are listed in Table 1. Those that are used less frequently in toxicity testing, but that are also valuable methods for mechanistic studies, are listed in Table 2.

Some criteria measure the net effect of different toxic actions and have the advantage of demonstrating a toxic end-point which may be common to many types of mechanisms. Such criteria include damage to the cell membrane or loss of viability. Other criteria are more selective and sensitive, but have the disadvantage of being less confident measurements of decreased cell viability. Examples of these include cloning capacity of cells, cytoplasmic ATP content, or specific enzyme induction. Inhibition of cell growth (also referred to as proliferation for the purposes of this discussion) is a sensitive indicator for cellular response to a chemical. When coupled with measurements of metabolism, the data can yield very convincing answers to mechanistic and toxicity testing problems. Simultaneous measurement of growth and viability are standard indicators of cell integrity, and together with the estimates of cytotoxicity derived from metabolic experiments, they can contribute significantly to the ability of the cell culture system to predict or assess toxicity. Cell proliferation,

TABLE 1
Some Common General Toxicity Criteria for Cell
Lines in Culture

Criteria	Methods for determining the degree of inhibition or injury
Cell morphology	Histological analysis using light microscopy
Cell growth	Cell counts; mitotic frequency; DNA synthesis
Cell division	Clone formation, plating efficiency
Cell metabolism	Anaerobic glycolysis; uptake of isotope-labeled precursors as indicators for newly synthesized macromolecules
Cell membrane	Leakage of isotope-labeled markers; leakage of enzymes (LDH, beta-glucuronidase)
Mitochondria	Mitochondrial integrity (MTT assay)
Lysosomes	Vital staining methods (neutral red assay)
Ribosomes	Synthesis of macromolecules (glycosamino-glycans); induction or inhibition of enzymes (acid and alkaline phosphatases)

TABLE 2
Less Frequently Used General Toxicity Criteria
for Cell Lines

Criteria	Methods for determining the degree of inhibition or injury
Cell morphology	Ultrastructural analysis using transmission electron microscopy
Cell growth	Mitotic frequency using karyotypic analysis
Total metabolism	Assays for specific proteins and hormones (collagen, steroids, neurotransmitters)
Cell membrane	Formation of blebs; uptake of trypan blue
Organelles	Isolation techniques for mitochondria, lysosomes, ribosomes
Cell staining	Immunohistochemical or cytochemical stains for proteins, enzymes, carbohydrates

however, is not reliable for demonstrating toxicity in cell lines which divide slowly, such as freshly prepared primary cultures or cells obtained from adult donors. Consequently, the criteria should include parameters which are either fully functioning or are unquestionably operational so as to form the basis of a dose response curve.

The first and most readily observed effect following the exposure of cells to toxicants is phase contrast morphological alteration in the cell layer and/or cell shape in monolayer culture. Therefore, it is not surprising that morphological alterations are used as an index of toxicity. Different types of toxic effects may also require investigative tools at different levels of sensitivity. Gross modifications such as blebbing or vacuolization are observed using light microscopy, whereas ultrastructural modifications require analysis by transmission or scanning electron microscopy.

Another indicator of toxicity is altered cell growth. The effect of chemicals on the ability of cells to replicate is used as an index of toxicity; the concentration of the substance at which 50% of the cells do not multiply is called the median inhibitory dose (ID_{50}). A more specific measure of replication is plating efficiency, the ability of cells to form colonies after 10 to 15 days in culture in the presence of a toxic agent. The information obtained with this parameter, together with cell growth, yields more complete information since both cell survival and ability to reproduce are monitored. Cell reproduction can be measured with several methods including cell count, DNA content, protein content, or enzyme activity. The assay for DNA content using biochemical methods or by monitoring the incorporation of radiolabeled precursors into DNA represents two different methods supplying similar information.

Another crude index of toxicity is cell viability. This endpoint can easily be measured by using vital stains such as trypan blue, which enters dead cells only, and neutral red uptake, which is actively absorbed by living cells. The latter is commonly used in biomaterial testing in the agar overlay method. A count of dead and vital cells is compared to the control and provides an index of lethality of the test compound. The release of ^{51}Cr is another index of lethality as a measure of membrane integrity (see Chapter 7).

Cells derived from different organs or tissues that retain some specialized functions *in vitro*, or that maintain specialized structures, are also widely used in toxicology. For these cells, effects on more specialized functions and/or structures are monitored in addition to effects on fundamental processes, including the monitoring of specific end products, metabolic pathways, and membrane integrity. Table 3 lists specific end points used for some cell types.

Because some cell systems do not possess an efficient metabolism, *in vitro* testing may require some form of metabolic activation. This is usually done in one of three ways:

1. By the addition of S9 fractions from rat liver; usually the mixed function oxidases are preinduced by treating the animals with phenobarbital, β-naphthaflavone, or Aroclor[2]
2. By preincubation of the test substance with a primary hepatocyte culture and addition of the preincubated medium to the test culture[3]
3. By coculture of the target cell with hepatocytes in the presence of the test substance.

In addition to the differences in toxic criteria, the cell line methods incorporate different incubation and exposure times and require varying media volumes, cell densities, serum concentrations, and gas phases. Traditionally, the earlier experiments used test tubes and glass petri dishes

TABLE 3
End-Points Commonly Used as Markers of Toxic Effects in Specialized Cells

Synthesis, release, or incorporation of specific molecules
Collagen mat, heme, hemoglobin, albumin, urea, lipoprotein, α-aminolevulinic acid, bile salts, metallothionein, glycosaminoglycans, incorporation of proline and conversion into hydroxproline, energy-dependent choline accumulation, histamine release, and cyclic AMP

Synthesis, activity, or release of specific enzymes
β-Glucuronidase, lactate dehydrogenase, ouabain insensitive ATPase, glucose-6-phosphate dehydrogenase, glycogen phosphorylase, glutamic-oxalacetic transaminase, glutamic-pyruvic transaminase, acetylcholinesterase, and renin

Interactions of compound with cells
Phagocytosis, cytoplasmic inclusions, intracellular accumulation, uptake or binding of compound to cytosol and lipoproteins, mitogenic response

Alterations of metabolic pathways
Hemoglobin reduction, glucose transport, 5-methyltetrahydropholate accumulation, hormone-stimulated gluconeogenesis, lipid peroxidation, fat accumulation, and glucosamine and galactose incorporation

Cell surface activities
Adhesiveness, conconavalin-A agglutination, antibody-mediated rosette formation, complement deposition on treated cell membrane, chemotactic migration, interaction with histidine H_1 receptor, GABA-mediated postsynaptic inhibition, spike frequency, membrane polarization, and electrophysiological alteration

which held media volumes of 5 or 10 ml. Recently, the methods have evolved to include 6-, 12-, 24-, and 96-well microtiter plates (the last type of plate has a capacity of 0.4 ml per well). This makes it possible to use small quantities of the test substances and permits for easier automation of the procedure. Thus the test solution can be diluted in serial or logarithmic steps using a microdiluter and many cultures can be analyzed simultaneously with a spectrophotometric plate scanner (Figure 1a and b). One critical variable during the course of an experiment is the exposure time. Intuitively, longer exposure times require lower doses of a chemical than the concentrations needed to cause toxicity for fewer hours. The toxicity for many substances therefore may increase 10- to 100-fold if the toxic end point is measured after 7 days rather than after 24 hours. Also, high serum concentrations tend to decrease toxicity of substances which are protein bound. The toxicity may also differ for a substance which is tested on cell monolayers, which have established intercellular contacts, as opposed to testing with freshly isolated cells.

The physicochemical properties of test compounds determine exposure conditions and the concentration of toxic agents in the culture medium. The medium is usually an aqueous saline solution with the addition of serum at

a

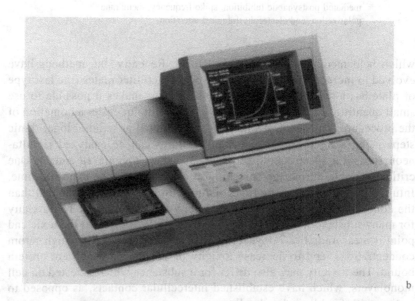

b

Figure 1. Photographs of (a) manual and (b) automated microplate readers featuring kinetic and endpoint assay analysis capability. Using 96-well plates, microplate applications for cytotoxicity testing include enzyme-linked immunosorbent assays (ELISA), protein and other colorimetric assays, cell counting, and enzyme kinetics. (Courtesy of BioTek Instruments, Winooski, VT.)

concentrations ranging from 2 to 15%. Only hydrophilic test compounds can be completely solubilized. For volatile toxic agents, special incubation steps are necessary to ensure a constant exposure with time, provided the partition coefficients and solubilities are known. Similar problems apply to hydrophobic test compounds or mineral particulates. Lipophilic substances can be solubilized in ethanol, methanol, or dimethylsulfoxide before addition to the medium; a control using the carrier solvent alone must be used. These practical considerations are discussed in Chapter 10.

In order to increase comparability of results and optimize testing procedures, standardization and adherence to the experimental approaches and procedures is desirable. The use of well-characterized cell lines, possibly of human origin, is important. Specific toxicity end points, as well as the most suitable assay methods, should be uniformly established. Cell cultures should be examined periodically for possible contamination with microorganisms, for cross contamination with other mammalian cell types, and for karyotypic integrity. Reports should provide details of exposure conditions, including information on the purity and source of the test compound. Moreover, the measurement of the concentration of the compound at the beginning and at the end of the experiment, particularly for insoluble or volatile substances, would facilitate the interlaboratory comparison of results. Two frequently used microtiter methods, the neutral red uptake assay and the MTT assay, are described. Both are suitable for rapid screening of growth and viability in continuous cell lines.

B. NEUTRAL RED UPTAKE ASSAY

The neutral red uptake cytotoxicity assay uses neutral red dye, which is preferentially absorbed into lysosomes.[4-8] The basis of the test is that a cytotoxic chemical, regardless of site or mechanism of action, can interfere with this process and result in a reduction of the number of viable cells. Only viable cells, therefore, are capable of maintaining the process intact. The degree of inhibition of growth, related to the concentration of the test compound, provides an indication of toxicity. Any chemical having a localized effect upon the lysosomes, however, will result in an artificially low reflection of cell number and viability. This factor does make the system useful for detecting chemicals which selectively affect lysosomes and thus more readily detects a specific mechanism of action. For example, chloroquin phosphate specifically alters lysosomal pH and thus has a greater effect on neutral red uptake than most chemicals.

For these experiments, cells are seeded in microtiter plates and allowed to settle and adhere for 24 hours (Figure 2). The medium is then replaced with fresh medium containing the test chemical at increasing concentrations. Three hours before the end of the exposure period, the medium is aspirated from the cells and replaced with neutral red solution (50 µg/ml in growth medium). The cultures are incubated for an additional 3 hours at 37°C. The medium is then

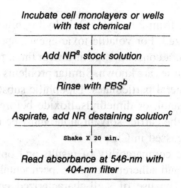

Figure 2. Outline of neutral red assay procedure. [a] Stock solution of neutral red (NR) is prepared as 5 mg/ml in Hank's buffered saline solution (HBSS) and added to incubating medium, to a final concentration of 50 μg/ml. [b] Phosphate buffered saline. [c] 1% glacial acetic acid, 50% ethanol in distilled water. (Adapted from Borenfreund and Puerner, *Toxicol. Lett.*, 24, 119, 1985.)

removed and the cells are rinsed with PBS, fixed, and destained with 1% acetic acid/50% ethanol. The plates are shaken for 10 min and the absorbance is read at 540 nm with a microplate reader, against a reference well containing no cells. Absorbances correlate linearly with cell number over a specific optical density range of 0.2 to 1.0.

One major drawback of the assay is the precipitation of the dye into visible, fine, needle-like crystals. When this occurs it is almost impossible to reverse and produces inaccurate readings. The precipitation is induced by some chemicals and thus a visual inspection stage during the procedure is important.

C. MTT ASSAY

In the MTT assay, the tetrazolium salt, 3-(4,5-dimethylthiazol-2-yl)-2,5-diphenyltetrazolium bromide (MTT), is actively absorbed into cells and reduced in a mitochondrial-dependent reaction to yield a formazan product.[9,10] The product accumulates within the cell since it cannot pass through the cell membrane. Upon addition of DMSO, isopropanol, or other suitable solvent, the product is solubilized and liberated and is readily quantified colorimetrically. The ability of the cells to reduce MTT provides an indication of mitochondrial integrity and activity which is interpreted as a measure of cell viability. The assay is suitable for a variety of cell lines displaying exponential growth in culture and a relatively high level of mitochondrial activity. It should be noted that some known compounds selectively affect mitochondria, resulting in an overestimation of toxicity.

The assay begins as with the neutral red procedure (Figure 3). During the final hour of incubation with the test chemical, a 0.5% MTT solution in growth medium replaces the test medium for 1 h at 37°C. The medium plus MTT is then replaced with DMSO and agitated for 5 min. The plates are read at 550 nm with a microplate reader, against a reference well containing no

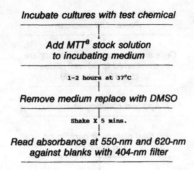

Figure 3. Outline of MTT assay procedure. [a] Stock solution is prepared as 5 mg/ml in HBSS. 10 µl of stock solution is added per well of 96-well plate. (Adapted from Borenfreund et al., *Toxicol. In Vitro*, 2, 1, 1988 and Mossman, *J. Immunol. Methods*, 65, 55, 1983.)

cells. Absorbances correlate linearly with cell number over a specific optical density range of 0.2 to 1.0.

Recently, the chemical XTT has been substituted for MTT, since the former is already soluble in aqueous media, and consequently, obviates the need for DMSO.

With both assays, the pH of the medium is quickly assessed by the change in color. Phenol red is used as the indicator in the medium and changes to a violet color when cells are totally inhibited, indicating a basic pH. With uninhibited cultures, a yellow-orange color develops. This occurs as a result of accumulation of acidic metabolites through anaerobic glycolytic pathways. With partially inhibited cells, the original red-orange color remains unchanged. Initial pH changes caused by test substances and precipitates of water-insoluble substances are also readily visible.

D. MIT-24 ASSAY

Recently, the MIT-24 procedure has been used for screening naturally occurring toxins present in food. The system measures acute cytotoxicity in sterile filtered supernatants of homogenized food, including perishable products, cooked foodstuff and ready-to-eat products. By using organ-specific cell lines, such as intestinal cells from mice, reliable and reproducible results have been gathered. This method has been routinely used for systematic testing of food additives.

The MIT-24 method has also been used in combination with chromatographic separation procedures to examine the toxic potential of components of food extracts, toxins present in shellfish and other seafood, and the glycoalkoloids which accumulate in potatoes. When the cellular method is combined with separation techniques such as high performance liquid chromatography (HPLC), low-level toxins can be measured which would otherwise remain undetected in animal tests. In addition, the MIT-24 system is useful in measuring the synergistic effects of combinations of chemicals.

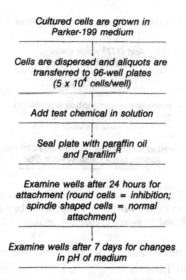

Cultured cells are grown in
Parker-199 medium

Cells are dispersed and aliquots are
transferred to 96-well plates
(5 x 10⁴ cells/well)

Add test chemical in solution

Seal plate with paraffin oil
and Parafilm^R

Examine wells after 24 hours for
attachment (round cells = inhibition;
spindle shaped cells = normal
attachment)

Examine wells after 7 days for changes
in pH of medium

Figure 4. Illustration of MIT-24 assay procedure. (Adapted from Ekwall, B., *Ann. N.Y. Acad. Sci.*, 407, 64, 1983.)

Figure 4 illustrates the MIT-24 procedure.[11] HeLa cells are cultured in Parker 199 medium with 10% fetal calf serum. Serial dilutions of the test chemical are prepared and added to each well of 24- or 96-well plates. Cells are suspended and rounded in a magnetic stirrer and added to the wells (50,000 cells per well). The plates are then sealed with mineral oil and Parafilm®. The test is based on the relative formation of rounded (metabolically inhibited) or spindle-shaped cells (not inhibited by the presence of the test chemical). In addition, the phenol red indicator in the medium reveals visual changes with the progression of metabolic activity. Basic pH (purple color) indicates totally inhibited cells, whereas acidic pH (yellow-orange color) demonstrates no inhibition due to accumulation of acidic metabolites from anaerobic glycolysis.

Other assays which have gained popularity for detection of general cytotoxicity include the FRAME kenacid blue method[12] and the pollen tube growth assay[13–15].

In general, validation studies have repeatedly shown that cytotoxicity tests with cell lines yield similar results, when the inhibitory concentrations are compared from experiments using identical incubation times. This indicates that the different cell types, as well as toxicity criteria, measure fundamental processes which allow comparison between otherwise dissimilar methods. The relatively small differences existing between cell lines probably depend on certain variations in their metabolic activity, leading to differences in the sensitivity of the experimental methods. Thus, the general toxicity illustrated by these test methods is identified as basal cytotoxicity. The suggestion is that any reasonable method which incorporates an established cell line will accurately measure basal cytotoxicity.

REFERENCES

1. **Schou, J. S.**, Mechanistic studies in man, laboratory animals and *in vitro* systems, *Food Chem. Toxic.*, 28, 767, 1990.
2. **Flint, O. P.**, An *in vitro* test for teratogens using cultures of rat embryo cells, in: *In Vitro Methods in Toxicology*, Atterwill, C. K. and Steele, C. E., Eds., Cambridge University Press, London, 1987, 339–364.
3. **Gomez-Lechon, M. J., Montoya, A., Lopez, P., Donato, T., Larrauri, A., and Castell, J. V.**, The potential use of cultured hepatocytes in predicting the hepatotoxicity of xenobiotics, *Xenobiotica*, 18, 725, 1988.
4. **Babich, H., Rosenberg, D. W., and Borenfreund, E.**, *In vitro* cytotoxicity studies with the fish hepatoma cell line, PLHC-1 (*Poeciiiopsis lucida*), *Ecotoxicol. Environ. Safety*, 21, 327, 1990.
5. **Babich, H., Shopsis, C., and Borenfreund, E.**, *In vitro* cytotoxicity testing of aquatic pollutants (cadmium, copper, zinc, nickel) using established fish cell lines, *Ecotoxicol. Environ. Safety*, 11, 91, 1986.
6. **Borenfreund, E. and Puerner, J. A.**, Toxicity determined *in vitro* by morphological alterations and neutral red absorption, *Toxicol. Lett.*, 24, 119, 1985.
7. **Borenfreund, E., Babich, H., and Martin-Alguacil, N.**, Comparisons of two *in vitro* cytotoxicity assays — the neutral red (NR) and tetrazolium (MTT) test, *Toxicol. In Vitro*, 2, 1, 1988.
8. **Ekwall, B.**, Screening of toxic compounds in mammalian cell cultures, *Ann. N.Y. Acad. Sci.*, 407, 64, 1983.
9. **Mosmann, T.**, Rapid colorimetric assay for cellular growth and survival: application to proliferation and cytotoxic assays, *J. Immunol. Methods*, 65, 55, 1983.
10. **Denizot, F. and Lang, R.**, Rapid colorimetric assay for cell growth and survival. Modifications to the tetrazolium dye procedure giving improved sensitivity and reliability, *J. Immunol. Methods*, 89, 271, 1986.
11. **Ekwall, B. and Acosta, D.**, *In vitro* comparative toxicity of selected drugs and chemicals in HeLa cells, Chang liver cells, and rat hepatocytes, *Drug Chem. Toxicol.*, 5, 229, 1987.
12. **Clothier, R. H., Hulme, L., Ahmed, A. B., Reeves, H. L., Smith, M., and Balls, M.**, *In vitro* cytotoxicity of 150 chemicals to 3T3-L1 cells assessed by the FRAME kenacid blue method, *ATLA*, 16, 84, 1988.
13. **Strube, K., Janke, D., Kappler, R., and Kristen, U.**, Toxicity of some herbicides to *in vitro* growing tobacco pollen tubes (the pollen test), *Environ. Exp. Bot.*, 31, 217, 1991.
14. **Kappler, R. and Kristen, U.**, Photometric quantification of *in vitro* pollen tube growth: a new method to determine the cytotoxicity of various environmental substances, *Environ. Exp. Bot.*, 27, 305, 1987.
15. **Kristen, U. and Kappler, R.**, The pollen tube growth test, in *Methods in Biology: In Vitro Toxicity Testing Protocols*, O'Hare, S., Ed., Humana Press, Hatfield Herts, U.K., 1992.

Chapter 4

CELLULAR METHODS OF TARGET
ORGAN TOXICITY

I. PRIMARY CULTURES OF SPECIALIZED CELLS

Homogeneous cell populations isolated in quantity from the framework of an organ present a unique opportunity to study the metabolism of a single cell type. There are a large number of systems described for primary cultures of specialized cell types which could be used for examining toxic effects of compounds on selected target organs. In general, the organs which have been of interest to toxicologists include the lung, liver, skin, kidney, hematopoetic, nervous and immune systems, and skeletal and cardiac muscle. With the exception of techniques which use hepatocytes or keratinocytes, standardized or validated test methods have not been developed to date from this resource of test methods. The reason for this is that primary cultures have several disadvantages for the assessment of organ specific cytotoxicity:

1. Animals are often needed for these cultures and the tests are more difficult to perform.

2. When used for detecting and quantifying systemic toxicity in target organs, primary cultures are limited to only a few of the many possible organ injuries which could be determined using animals. This is a fundamental limitation of the technical aspects of establishing primary cultures. In conventional whole animal experiments, any or all of the organs which may or may not show gross pathology can be removed at the time of sacrifice. The organs can then be frozen or preserved indefinitely with fixatives according to the analytical protocol. Testing chemicals with potential to produce local irritation, however, has been routinely carried out using primary cultures and early subcultures of keratinocytes for skin irritation, corneal cells for eye irritation, and gingival cells for testing dental materials.

3. In the past, primary cultures were difficult to standardize because the cells were derived from different animals. This is no longer valid since animal breeding has developed into a very refined science. The problem, however, still exists when human donors are selected. Even when efforts are made to match the ages and sex of the donors, a primary culture from one individual is not necessarily identical to another. This impediment can be overcome by performing the necessary basic tests to characterize a primary culture, including karyotypic analysis and identification of functional markers. By including this battery of procedures in the analytical method, the progress of the experiments is facilitated and interlaboratory comparisons are valid.

Despite the disadvantages of primary cultures, many types of specialized cells have been used to detect the target organ toxicity of chemicals. Often, a prototype chemical and its analogues are tested so that their toxicity could be compared in the same system. This permits established toxicity data for some of the substances to be used as a reference for previously unknown toxicity of unrelated materials. Hepatocytes have been the subject of this testing principally because of the important role that the liver demonstrates in xenobiotic metabolism, i.e., metabolism of exogenous substances. Other cell types that have been traditionally employed include lung macrophages and epithelial cells, blood cells, reticuloendothelial cells, cardiac cells, renal cells, neuronal cells, and muscle cells. Procedures describing the isolation of cells from liver, kidney, and lung are detailed below, while studies using cell culture methods for detecting chemicals which cause local toxicity or affect the immune system are described in later chapters. In addition, Tables 1 through 6 summarize some enzymatic conditions which are currently used to dissociate various tissues.

A. ESTABLISHMENT OF PRIMARY CULTURES
OF HEPATOCYTES

In humans the liver plays an important role in drug and xenobiotic metabolism. Following administration of a drug or chemical, the agent is screened by the liver. The *first pass* phenomenon occurs immediately after oral absorption, as the circulation of the upper intestinal tract is intimately associated with that of the biliary tracts. Thus, exposure of an agent to the liver renders it susceptible to biochemically and physiologically important metabolic processes. These processes are termed pharmacokinetics, i.e., the qualitative and quantitative study of the time course of absorption, distribution, biotransformation, and elimination of an agent in an intact organism or cellular system. Toxicokinetics refers to the identical processes which are characterized for a substance which has the potential for toxicity. Thus the study of toxicokinetics can yield useful information about the disposition of a toxic compound following administration. This information can be used to predict or monitor the metabolic products of the parent compound, whether toxic or nontoxic, the duration of the effect, the sequestration into body compartments, and the route of elimination.

The process of metabolizing or detoxifying a chemical *in vivo* usually follows a two-step mechanism which, in combination, results in the formation of a water soluble, nontoxic product.[1-4] The first step in the process, *phase I biotransformation* reactions, involves a group of cellular enzymes, including monooxygenases, esterases, and hydrolases, which are collectively known as the *cytochrome P450* proteins. Through a series of reduction-oxidation reactions, the active sites of chemicals are converted to expose electrophilic, polar substituents. At the completion of these reactions, *phase II biotransformation* reactions proceed to couple normal plasma and interstitial constituents to the electrophilic sites of the metabolite. These groups consist of acetyl, glucuronic

TABLE 1
Enzymatic Conditions Used to Dissociate Liver Tissue

Species and age	Dissociating enzyme	Medium[a]
Fetal rat	Collagenase 0.05 to 0.1% Hyaluronidase 0.1% DNase 0.00125%	Ca/Mg-free HBSS
	or	
	Collagenase 0.05% DNase 0.00125%	Ca/Mg-free HBSS
Neonatal rat	Crude trypsin preparations 0.25%	Ca/Mg-free PBS
Young adult rat[b]	Collagenase 0.5% Hyaluronidase 0.1%	Ca/Mg-free HBSS
Older adult rat[c]	Collagenase 0.05% Hyaluronidase 0.1%	Ca/Mg-free HBSS
Adult mouse[d]	Collagenase 0.05% Hyaluronidase 0.1%	Ca/Mg-free HBSS

[a]   Hank's balanced salt solution, phosphate buffered saline.
[b]   50 to 150 g.
[c]   150 to 300 g.
[d]   20 to 75 g.

TABLE 2
Enzymatic Conditions Used to Dissociate Muscle Cells

Species and age	Dissociating enzyme	Medium[a]
Fetal rat	Collagenase 0.5%	Ca/Mg-free HBSS
Neonatal rat	Crude trypsin preparations 0.125%	Ca/Mg-free PBS
	or	
	Trypsin 0.006% Chymotrypsin 0.0015% Elastase 0.0025%	
Young adult rat[b]	Collagenase 0.1% Hyaluronidase 0.05%	Ca/Mg-free HBSS
Older adult rat[c]	Collagenase 0.1%	Ca/Mg-free HBSS
	or	
	Collagenase 0.1% Hyaluronidase 0.2%	

[a]   Hank's balanced salt solution, phosphate buffered saline.
[b]   50 to 150 g.
[c]   150 to 300 g.

TABLE 3
Enzymatic Conditions Used to Dissociate Endothelial Cells

Species and age	Dissociating enzyme	Medium[a]
Bovine, adult pulmonary artery aorta	Collagenase 0.05 to 0.25%	Ca/Mg-free HBSS or Ca/Mg-free PBS
Rodent, adult[b] lung	Collagenase 0.05 to 0.25%	Ca/Mg-free HBSS
Rabbit, adult[c] lung	Collagenase 0.05 to 0.25%	Ca/Mg-free HBSS
Bovine, young adults pulmonary artery microvessels	Collagenase 0.05 to 0.25% Scraping	Ca/Mg-free HBSS

[a]   Hank's balanced salt solution, phosphate buffered saline;
[b]   50 to 150 g;
[c]   1.5 to 2.0 kg.

TABLE 4
Enzymatic Conditions Used to Dissociate Lung Tissue

Species and age	Dissociating enzyme	Medium[a]
Fetal rat	Collagenase 0.5%	Ca/Mg-free HBSS
Fetal rabbit	Trypsin 0.25% Elastase 0.05% DNase 0.003%	Ca/Mg-free HBSS
Adult rat[b]	Collagenase 0.1% or Collagenase 0.1% Crude trypsin preparations	Ca/Mg-free PBS
Adult rabbit[c]	Collagenase 0.1% Pronase 0.2% Elastase 0.05% DNase 0.003% or Trypsin 0.25% Elastase 0.05% DNase 0.003%	Ca/Mg-free HBSS

[a]   Hank's balanced salt solution, phosphate buffered saline.
[b]   50 to 150 g.
[c]   1.5 to 2.0 kg.

acid, and glutathione moieties which convert the metabolites to water-soluble compounds. These nontoxic products are then ready for excretion into the urine or feces.[5] It is important to note that although the majority of substances are detoxified and inactivated through these reactions, some substances are activated from nontoxic or inactive parent compounds to active toxic metabolites.

TABLE 5
Enzymatic Conditions Used to Dissociate Neural Tissue

Species and age	Dissociating enzyme	Medium[a]
Rat brain 15 to 16 day fetus (brain reaggregates)	Mechanical dissociation	Ca/Mg-free HBSS or Hank's D2
Mouse newborn 0 to 30 days	Trypsin 0.25%	HBSS
Rat, calf, lamb unspecified	Trypsin 0.1 to 1.0%	HBSS

[a] Hank's balanced salt solution.

TABLE 6
Enzymatic Conditions Used to Dissociate Thyroid Tissue

Species and age	Dissociating enzyme	Medium[a]
Porcine adult	Trypsin/collagenase 0.25%/0.05%	Ca/Mg-free HBSS
Bovine, porcine	Trypsin 0.004%/ Collagenase 0.0025%	Earle's BSS
Chicken embryo	Collagenase 0.2%/ Trypsin 0.2%	Ca/Mg-free Tyrode's solution
Chicken embryo	Collagenase 0.25%	Tyrode's saline
Chicken embryo	Collagenase 0.25%/ ± Trypsin 0.1%	Ca/Mg-free HBSS plus chick serum

[a] Hank's balanced salt solution.

The fact that the liver has the highest capacity of all organs to alter chemical substances does not preclude the ability of other organs to contribute to the toxicokinetics of a compound. The rationale for using hepatocyte cultures, however, is based on the abundant physiological potential of the liver for biotransformation *in vivo*. In addition, the assumption is also implied that a drug which exerts an effect on the intact organ will also act in a similar fashion on the isolated cells. Based on this premise, many of the initial studies using hepatocyte cultures presented conclusions for selective target organ toxicity of the chemical.[6,7] Admittedly, target organ toxicity can be correlated with the ability of the liver from a particular species to biotransform an agent and effectively detoxify it. The extrapolation from *in vitro* to the *in vivo* situation, however, does not depend exclusively on the final products of metabolism nor on the selective role of the liver to alter the chemical. In fact, target organ toxicity may be the expression of basal cytotoxicity that occurs as a result of accumulation of the compound in the liver (see Chapter 2, Section II.)

Hepatocyte cultures are also suitable for use as screening methods for systemic toxicity, since these cells can provide information on basal, as well as organ-specific, cytotoxicity.[8] By virtue of the extensive metabolic capacity of the liver to screen most circulating xenobiotics, hepatocytes have the propensity to screen toxicants which are selectively altered or sequestered by the liver. A serious disadvantage to the use of hepatocytes, however, is that they cannot be routinely maintained as differentiated continuous cultures and must be established as primary cultures for each experiment. Some of the various methods used for the isolation and maintenance of primary cultures of hepatocytes include *in situ liver perfusion and digestion, primary hepatocyte enrichment, and hepatocyte cell culture* (suspension and attachment cultures).

1. *In Situ* Liver Perfusion

In situ liver perfusion can be performed routinely once the initial setup is in place.[9,10] Essentially, the procedure relies on the pumping of three separate fluids through the intact rat liver and collecting the cells at the end of the last wash step. The first step involves the pumping of the wash perfusion medium, formulated in HBSS, containing heparin as the active ingredient. The solution removes blood from the portal circulation and is infused through the vena cava at 25 to 35 ml/min through one arm of a "Y" tube connector. After 10 min, the digestion perfusion medium, also formulated with HBSS and containing collagenase as the active component, is infused through the other arm of the "Y" connector for 10 min. The liver is then transferred to a petri dish and 20 ml of a fresh cell wash medium is injected into the organ. Further release of hepatocytes is stimulated by mechanically agitating the dish, making small incisions in the organ, and scraping the lobes of the liver. The procedure is repeated and the solutions are collected and centrifuged. Primary cultures are initiated by inoculating culture flasks with the cell suspensions. All solutions are prepared, and manipulations are performed, under aseptic conditions.

Perfusion of explanted liver specimens from a human or rodent requires aseptic technique in a manner not unlike that described for *in situ* perfusion. Briefly, biopsied material is perfused through the exposed blood vessels on the cut surfaces of the organ. Perfusion, digestion, and wash-out media are similar in composition as above, and flow rates are maintained constant and recirculated. Primary hepatocytes are released by gentle combing after transferring to sterile petri dishes.

2. Primary Hepatocyte Enrichment

Suspensions of isolated liver cells may be manipulated so as to remove nonhepatocyte and nonviable cells from the desired primary hepatocytes. A low-speed, isodensity Percoll® centrifugation procedure used to enrich the population of primary hepatocytes from liver cell isolates has been described by Kreamer et al.[11] Using an aseptic technique, a suspension of freshly isolated and washed cells is layered and then mixed with a stock solution of isoosmotic Percoll® and centrifuged at $50 \times g$ for 10 min at 4°C. The parenchymal hepatocytes are separated and removed from the bottom of the tubes while contaminating and nonviable cells float on top. The primary cells are then resuspended in appropriate medium in preparation for primary cultures.

3. Hepatocyte Cell Culture

As with maintenance of most cell cultures, freshly isolated hepatocytes can be established as suspension or attachment cultures. The criteria for selecting one system or the other is based on the degree of differentiation that the primary cultures undergo once they are exposed to the artificial growth conditions and the consequent loss of differentiated functions inherited and carried over to some extent from their *in vivo* existence. A fundamental discussion of the differences between primary and continuous cultures is found in previous sections (Chapter 2, Sections III and IV). For primary cultures of parenchymal liver cells in particular, suspension cultures are not desirable. Suspension cultures of these cells result in a rapid loss of differentiated functions and cytochrome P450 levels. Thus cytotoxicity tests must be performed generally within 12 h.[6]

Attachment cultures using basement membrane substrata are more suitable. Establishing primary cultures on plastic cell culture plates coated with cellular attachment components maintains differentiated features, improves viability and membrane integrity, especially after the vigorous isolation procedures, and preserves cytochrome P450 for several weeks. The substrata include collagen types I and IV, laminin, fibronectin, or combinations of these, are applied as solutions to the culture plate surfaces, and are commercially available or can be prepared in the laboratory.

4. Functional Markers of Primary Hepatocyte Cultures

Table 7 summarizes some of the important criteria used for assessing functional and differentiated features of freshly isolated liver parenchymal cells.[12] These criteria are also incorporated in cytotoxicity testing of chemicals which exhibit organ-specific toxicity. In fact, one or more test methods may be used to screen chemicals for potential liver toxicity. In addition, the ability of a chemical to undergo biotransformation is best evaluated by using primary cultures of hepatocytes.

B. ESTABLISHMENT OF PRIMARY CULTURES OF RENAL CELLS

1. Isolation of Renal Cortical Cells

Several methods have now been thoroughly investigated which effectively yield sufficient quantities and purity of cortical renal cells from a variety of animal species.[13-18] The cortical cells stem from the renal cortex and constitute the most metabolically active cells of the kidney. Of these cells, the ones most frequently used for the assessment of renal toxicity include the proximal tubule and distal tubule cells. The methods of isolating and obtaining enriched populations

TABLE 7
Cytotoxicity Criteria of Primary
Hepatocyte Cultures

Cell Membrane Integrity
Trypan blue exclusion
Lactate dehydrogenase leakage
Aspartate aminotransferase leakage

Phase I Metabolic Reactions
7-Ethoxycoumarin-*O*-deethylation activity

Phase II Conjugation Reactions
Glutathione Determination
Glutathione *S*-transferase Activity

Peroxisomal β-Oxidation
Cyanide insensitive acyl-CoA oxidase
Carnitine acyltransferase

of these cells rely on the ability of the procedures to selectively screen out other contaminating cell types, such as cells associated with the glomerulus and cortical collecting duct.

The most frequently employed protocols include enzymatic methods, mechanical methods, and historically, microdissection techniques. Any of these may be coupled with density gradient centrifugation on Percoll® to enrich the fractions for proximal or distal tubule cells.

Although developed as one of the original methods for isolating cortical cells, *microdissection* has limited use because of the poor yields of tubule cells and the prolonged times required for the procedure. Essentially, the method requires precise use of microdissecting capability to focus on nephron tubules. The method is laborious and, for the purposes of screening nephrotoxic chemicals, does not afford enough cells for a suitable cytoxicity testing procedure.

Mild enzymatic digestion with purified collagenase in a balanced salt medium is the method of choice for releasing cortical cells from the basement membrane. In general, the kidneys from a surgically anesthetized rabbit or rodent are first perfused *in situ* with Hank's balanced salt solution (HBSS). Slices of kidney cortex are then digested in mild collagenase suspensions for 1 or 2 h at temperatures ranging from 4 to 37°C (alternatively, the kidneys are removed and perfused with the collagenase suspension). The initial homogenate contains tubule cells which have been dislodged from the tubular basement membrane, as well as aggregates of undigested tubules, undesirable segments of the nephron, intact nephron structures, and connective tissue. This homogenate is then passed through a syringe filter (200 to 500 μm mesh) or nylon screening sieve. The cell suspension is further enriched by layering on a continuous Percoll® density gradient (10 to 55%) and centrifuged at low speed, after which the proximal tubule cells are aspirated from the bottom of the tubes.

Mechanical separation of proximal cells relies on the *in situ* perfusion of the kidney with a suspension of iron oxide particles. These micronized pellets lodge in the glomeruli and allow for the separation of glomeruli from proximal tubule cells. That is, after perfusion, portions of the cortex are hand homogenized, coarsely filtered, and the suspension is exposed to a mechanical stirring bar to which the iron-laden glomeruli are attracted and separated.

Thus, selection of a particular isolation procedure must primarily yield a sufficient number of cells adequate for toxicity testing and, in addition, the cells must be of sufficient purity so that fundamental conclusions can be drawn with confidence. Underlying this premise is the need to assess the functional characteristics of the cell types. These same characteristics, which are described below, can be used as markers for cytotoxicity testing during the course of the screening protocols.

Subsequent culture of the isolated tubules and their corresponding cells can be accomplished on plastic surfaces of tissue culture flasks or wells, on membrane filter inserts, or on biomatrix-coated plastic surfaces as described in this chapter for hepatocytes (Section I.A). Proximal and distal cells generally migrate from their respective isolated tubules within 24 to 48 h and reach stationary levels in approximately 1 or 2 weeks. A variety of basal growth media have been used, such as Dulbecco's modified Eagle's medium and Ham's nutrient mix F12; both serum-free and serum-supplemented.

2. Use of Continuous Renal Cell Lines

Over the last 20 years, improved methods for maintaining continuous cultures of cells have allowed for the realization of general mechanistic cytotoxicity studies. Renal cell culture technology has benefited primarily from the improved technology in much the same way as in other fields. The problems associated with other continuous cell lines, however, have also plagued the maintenance of continuous cultures of renal cells. These problems include those which are discussed above (Chapter 2, Section IV), the most important of which include the loss of *in vivo* characteristics following isolation and subsequent passage over several generations *in vitro*. This dedifferentiation may be followed by a decrease in the levels of functional markers necessary for identifying the cell type and for assessing cytotoxicity.[18] Nevertheless, the propagation of continuous cell lines has allowed for important contributions to understanding renal cell biology and physiology. It may be necessary to approach renal cytotoxicology, however, more cautiously when using cell lines. As with the use of continuous cell lines from any organ, the apparent toxicity from a chemical may be a reflection of basal cytotoxicity to the cell rather than a specific renal toxic effect.

Among the most frequently used permanent cell lines used in renal studies are the transformed cells which have been in the laboratory for decades, including the MDCK (Madin-Darby canine kidney) and LLC-PK$_1$ renal epithelial cell lines.[19,20]

MDCK cells maintain morphological features and functional characteristics consistent with cells of distal tubule/collecting duct origin. In addition, they show biochemical evidence of transporting epithelia, although several strains and clones have been established which differ considerably from the parent cell line and from each other. LLC-PK$_1$ cells are derived from pig proximal tubule cells and share morphological features characteristic of their parent cells, including the ability to form domes, the presence of apical membrane microvilli, and several junctional complexes. Physiologically, they express high activities of brush border membrane marker enzymes and retain the ability to transport small molecules.

3. Characterization of Renal Cells

Table 8 summarizes important criteria used for assessing differentiated morphological and functional features of primary and continuous cultures of renal cortical cells. These features provide evidence for the origin of the cells and lend support during the course of investigations, for the continuing reflection of the properties of the parent cells *in vivo*. Many of these methods, alone or in combination, have also been used to evaluate the cytotoxic potential of nephrotoxic chemicals. The simplest procedures, and consequently the ones most frequently employed, measure leakage of marker enzymes, which are indicative of membrane damage. Also, a unique quality of these cells is the transport function of the cultured epithelia, including the monitoring of the passage of molecules across the epithelial layer, as well as the change in electrophysiological measurements occurring at the stationary phase of culture. Finally, periodic monitoring of the presence or absence of ultrastructural features known to be present in the parent cells can be used to assess the purity of the differentiated cells in culture.[21-24]

C. ESTABLISHMENT OF PRIMARY CULTURES OF LUNG CELLS

Mechanistic studies of organ-specific toxicity in the mammalian lung have been hampered by the heterogeneity of its composition. Over 40 different cell types have been identified in this organ, not all of which have been extensively characterized. This complex cellular organization, together with their interaction with the extensive vascular, lymphatic, and neuronal networks, has presented formidable challenges to the performance and interpretation of mechanistic studies on lung-specific toxicity. Because of the numerous cell types, only a few have been isolated, purified, and characterized which possess considerable biomedical research interest.

The prospect of isolating and maintaining relatively pure primary cultures of select lung airway cells is technically challenging. Only recently has it been possible to separate pure populations of epithelial cells from the mammalian lung. Initial techniques were devised which used enzymatic digestion of perfused, lavaged, or minced rodent or rabbit lung.[25] Various improvements have

TABLE 8
Cytotoxicity Criteria of Primary Cortical Cell Cultures

Parameter	Metabolic/target site	Type of injury
Trypan blue exclusion	General viability, tubule transport	Metabolic, proliferation
Cell growth	General growth, viability	Metabolic, proliferation
Total protein	General growth, viability	Proliferation, energy depletion
Enzymatic		
LDH release	Cytoplasmic membrane	Membrane integrity
MTT reduction	Mitochondrial reduction	Oxidative
ATP content	Mitochondrial reduction	Energy depletion
NR uptake	Lysosomal uptake	Transport/storage
GSH	General metabolic processes	Oxidative
DNA	Nuclear	Electrophilic attack
GGT, AP, SD, ATPase	Proximal tubules, brush borders	Oxidative

incorporated centrifugal elutriation, laser flow cytometry, or unit gravity sedimentation.[26-36] Continuous culture of cell strains has generally resulted in the disappearance of characteristic features and loss of cell proliferation soon after initial isolation. Table 9 lists some of the major cell types currently used in pulmonary studies, along with their unique biochemical features. These cells have also found utility in cytotoxicological studies.

1. Isolation and Culture of Epithelial Cells

The isolation and maintenance of relatively pure populations of alveolar epithelial cell lines has met with considerable resistance. Only recently has it been possible to separate alveolar type II cells from the mammalian lung. Kikkawa and Yoneda[25] initially devised a technique for separating epithelial cells from rat lung. Various improvements have since appeared which incorporate enzymatic digestion of normal rodent lung followed by centrifugal elutriation or laser flow cytometry.[28,29] Today, type II pneumocytes have been maintained in continuous culture following isolation from rat, rabbit, cat, hamster, and mouse. They retain high levels of enzymatic activity, display cytoplasmic lamellar inclusions, and synthesize and secrete surfactant. These features are maintained in culture for many weeks and, as a result, have been used in mechanistic studies of oxygen toxicity,[37] in the study of airborne pollutants such as trichloroethylene, carbon tetrachloride, tobacco smoke,[38] and with chemicals that have an affinity for alveolar epithelial cells, such as paraquat and taurine.[39,40] Continuous culture of cell strains, however, has generally resulted in the disappearance of characteristic features, such as the

TABLE 9
Cell Types Isolated from the Mammalian Lung

Cell type	Derivation[a]	Biochemical features
Type II pneumocytes	Alveolar epithelium	Manufacture and store pulmonary surfactant; xeno-biotic metabolism
Clara cell	Non-ciliated bron-chiolar epithelial cell	Xenobiotic metabolism
Ciliated epithelial cells	Bronchiolar, tracheal origin	Tissue repair, cellular protection
Endothelial cells	Capillary, venous, arterial origin	Gas exchange, solute transport, metabolic secretion, diaped-esis of immune cells
Macrophages	Alveolar origin	Phagocytic proper-ties, immune activation

[a] Cell type from which primary cells were originally isolated.

loss of cytochrome P450 enzyme activity, and reduction in proliferation within 2 weeks of initial isolation.[32]

A typical protocol used in the isolation of alveolar epithelial cells is described. All procedures involving isolation and culture of cells are performed aseptically. Young adult male rabbits are anesthetized with pentobarbital (50 mg/kg intravenously). Airway pressure is maintained constant (10 cm H_2O, 25°C) with N_2 by tracheal cannulation. Lungs from anesthetized animals are surgically removed, trimmed, rinsed, cut into 1 mm^3 pieces, and placed in culture bottles containing Hank's balanced salt solution (HBSS) plus 1.25 mg/ml trypsin, 0.3 mg/ml elastase, and 0.015 mg/ml DNAse. The mixture is stirred mechanically for 15 min at 37°C, the suspension is filtered through gauze and the remaining undigested lung is reincubated with fresh protease solution. An equal volume of medium containing fetal bovine serum (10%) is added to the filtrate, containing the dispersed cells, to neutralize the enzyme reaction. The procedure is repeated three times; the filtrates are combined, pelleted, and the cell pellet is resuspended and incubated at 37°C in Dulbecco's modified Eagle's medium (DMEM) using differential adhesion (this manipulation is based on unit gravity sedimentation which separates cells by size and density; fibroblasts, the main contaminant in primary cultures, settle faster than epithelial cells, thus the cell suspension is plated for one hour after which the medium is replated in separate flasks). DMEM is supplemented with 10% fetal bovine serum and the gas phase is 10% CO_2/90% air. Figures 1, 2, and 3 show phase contrast micrographs of isolated lung cells in culture at different microscopic magnifications.

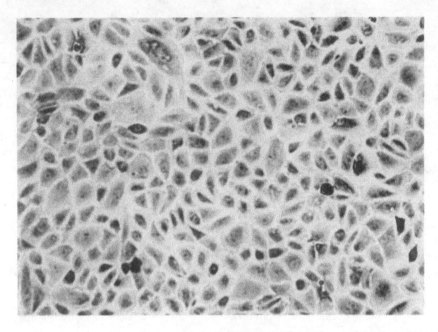

Figure 1. Phase contrast micrograph of alveolar epithelial cells in culture isolated from rabbit lung. Cells were fixed with neutral buffered formalin 10% and stained with 1% toluidine blue in 70% ethanol. The cobblestone arrangement is characteristic of lung epithelial cells in culture. (Magnification × 100.)

Primary cultures and subcultures are routinely monitored for viability using trypan blue exclusion, growth, and homogeneity using phase contrast microscopy. Alternatively, the Papanicoulou stain is also used for monitoring the monolayer since the stain is specific for lipid inclusions. These cells are rich in surfactant, which is stored in cytoplasmic lamellar inclusions. For subculturing, confluent monolayers are rinsed with phosphate buffered saline containing sodium bicarbonate and are then exposed to 0.125% trypsin solution. Cells are passaged and population doubling levels (pdl) are calculated according to routine cell culture procedures.

Organotypic cultures represent a model system suitable for the investigation of *in vitro* cytotoxic responses. The description of the protocol for establishing this system was originally described by Douglas et al.[41] and modified by others.[37] The cells, generally isolated from fetal, newborn, and postnatal rats, retain differentiated characteristics for longer periods of time in culture and maintain their capacity to produce and secrete pulmonary surfactant. They also reaggregate and form "alveolar-like" structures when layered on Gelfoam® squares, thus resembling their *in vivo* appearance (Figure 4). Surfactant synthesis was suggested by the appearance of lamellar inclusions in the cytoplasm and lamellar structures in the extracellular space. Organotypic cultures have also been established from fetal lungs of the human, mouse, rabbit, lamb, and monkey.[37] Consequently, they are useful in screening the effects of various

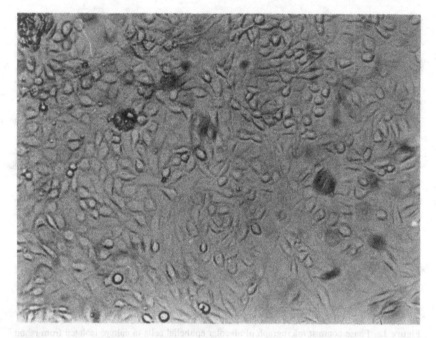

Figure 2a. Phase contrast micrograph of alveolar epithelial cells in culture isolated from rabbit lung. Photographed using Nomarski optics showing the domed cuboidal morphology of the cells and some mitotically active figures. (Magnification × 100.)

regimens which may be beneficial in treating the pathology of immature mammalian lungs. In cytotoxicology, organotypic models of type II pneumocytes are useful for identifying agents with potential capacity for oxidant injury.

Another reproducible method for the separation of alveolar epithelial cells in particular, and for purification of specific subpopulations of cells from heterogeneous mixtures of cell suspensions in general, lies in *unit gravity sedimentation*. The procedure relies on the separation of subpopulation of cells based on sedimentation rate at unit gravity and is determined primarily by the size of the cell. Essentially, tissue from a target organ is isolated, enzymatically digested according to procedures similar to those described above, and loaded onto a Ficoll® linear density gradient solution. The chamber is then oriented to the horizontal position, and the zones of interest containing the separated cells are collected as fractions. The cells are subsequently collected, analyzed, and cultured as desired. The method has been applied to several target organs and their respective population of cells with unique functions that are amenable to separation with this technique.[35]

2. Isolation and Culture of Clara Cells

The lining of the distal airways of the lungs consists mainly of ciliated and non-ciliated cells. The Clara cell is an example of a nonciliated bronchiolar epithelial cell originally identified as having a secretory role. The

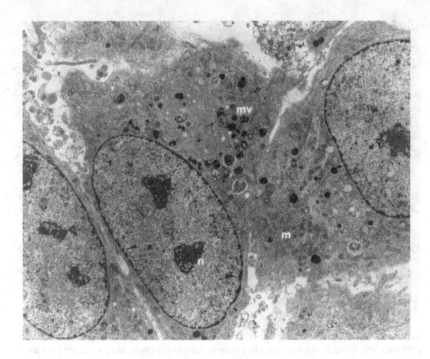

Figure 2b. Electron micrograph of alveolar epithelial cells in culture isolated from adult rabbit lung displaying numerous electron opaque multivesicular bodies (mv), lamellar-like inclusions (open arrow), and mitochondria (m). These mitotically active cells also demonstrate condensed nucleoli (n) and limited amount of heterochromatin. (Arrowheads, magnification × 7000.) (From Houser, S. et al., *ATLA*, 17, 301, 1990. With permission.)

nature of the putative secretory product and the mechanism of secretion has not been fully established. The cells have been reported to stain for phospholipid,[42] although they are distinguished from type II epithelial cells in primary culture by their lack of reactivity with the Papanicolaou stain,[25] and for alkaline phosphatase.[43,47] Clara cells in monolayer cultures have been identified according to their reaction with the nitroblue tetrazolium procedure.[32]

Ultrastructurally, as with pulmonary endothelial cells (Figures 5 and 6), the presence of an extensive endoplasmic reticulum, abundant mitochondria, and numerous opaque osmiophilic granules, suggest high metabolic activity. In fact, high levels of cytochrome P450 monooxygenase isozymes have been identified in Clara cells in several studies involving immunohistochemical and autoradiographic preparations, and in isolated cells[30,48-51] (Figures 7a and 7b).

The fact that metabolically active cells are targets for chemical injury may be related to the presence of xenobiotic metabolizing enzymes. Both Clara cells and type II pneumocytes have been shown to play a major role in detoxification

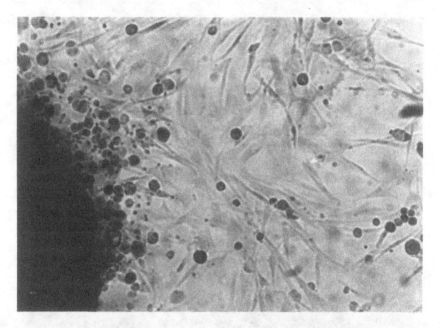

Figure 3. Phase contrast micrograph of fibroblasts in explant culture. The dark contrast of the explanted lung tissue shows rounded cells emerging from its periphery, after which the phenotypic appearance of the cells transforms to a spindle shape. (Magnification × 100.)

of xenobiotics, as a result of their enzymatic capacity. Devereux et al.[44] demonstrated that the pulmonary toxin, 4-ipomeanol, is activated by isolated alveolar pulmonary type II cells and Clara cells. The greater amount of metabolism in Clara cell preparations and the limited amount of damage to type II cells *in vivo* suggests that the former is the specific target for this chemical.[52] Furthermore, Clara cells have demonstrated a propensity for target cell interactions for a number of pulmonary toxicants, such as butylated hydroxytoluene[45] and 3-methylfuran.[46]

Clara cells are isolated along with alveolar type II pneumocytes in the same preparation using centrifugal elutriation as outlined in Figure 7c.[34,44,47] In general, most of the alveolar macrophages are removed during lavage since alveolar macrophages and Clara cells are similar in size and density. The main advantages of centrifugal elutriation are speed and separation of large numbers of cells. Type II cells are efficiently separated with the density gradients because of their uniform size and low density. Conversely, Clara cells are more difficult to isolate because of the variable density, small size, and numbers in the lung. Total cell yields are improved using the Percoll® density gradient fractionation and second elutriation.

3. Isolation and Culture of Alveolar Macrophages

The isolation and culture of pulmonary alveolar macrophages is described in Chapter 7.

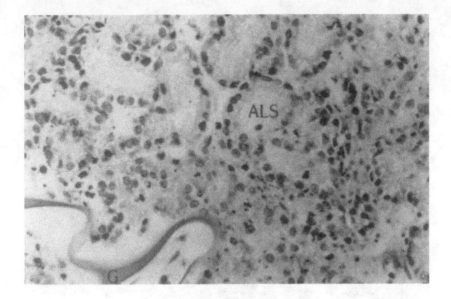

Figure 4. Organotypic cultures from fetal rat lungs of 18 to 19 days gestation. Cultures were grown in Ham's F-12 medium in an atmosphere of 10% oxygen, 5% carbon dioxide. Many more cells are seen in these cultures when compared to those of more mature animals. ALS, alveolar-like structure; G, gelfoam. (Magnification × 960.) (From Loo et al., *Exp. Lung Res.,* 15, 597, 1989. With permission.)

4. Isolation and Culture of Endothelial Cells

The pulmonary alveolar-capillary network lies beneath the basement membrane and comprises about 20% of the total body capillary system. These vessels are lined with fenestrated and non-fenestrated endothelial cells which function in gas exchange and in the movement of low and high molecular weight solutes. The cells regulate the transport into the circulation of important chemical mediators such as biogenic amines, kinins, angiotensins, and prostaglandins. In addition, they monitor diapedesis of monocytes, granulocytes, neutrophils, and lymphocytes.[53,54]

The pulmonary endothelium is ideally suited for regulating the entry of vasoactive substances into the systemic circulation. Endothelial cells of the lungs are known to conduct a variety of specific metabolic activities important in the pulmonary processing of vasoactive substances. Enzymes, chemically mediated inhibitors, cytoplasmic receptors, and transport systems of endothelial cells determine the level of hormones entering the circulation. It is by the regulation of these mechanisms that the pulmonary endothelium plays an important role in the response of the lung, as a target organ, to xenobiotics.

Endothelial cells from large and small vessels of the lung possess specialized but different properties, which have hindered the isolation of pure populations. The presence of fenestrations in cells from capillary beds, the appearance

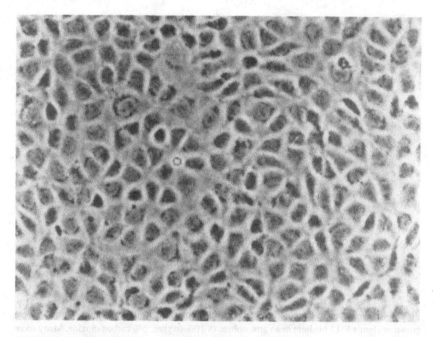

Figure 5. Phase contrast micrograph of a confluent monolayer of bovine pulmonary artery endothelial cells in their 12th passage. The "cobblestone" morphology is a requirement for characterizing these cells in culture. (Magnification × 250.) (From Ryan, U.S., *Environ. Health Perspect.*, 56, 103, 1984. With permission.)

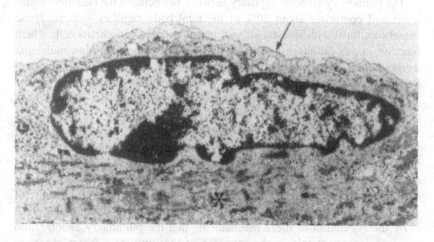

Figure 6. Electron micrograph of bovine pulmonary artery endothelial cells freshly isolated by scraping with a scalpel. In addition to caveolae on the luminal surface (arrow), Golgi apparatus (G) is apparent. The cytoplasm is replete with rough endoplasmic reticulum and fibrillar material (*). (Magnification × 20,000.) (From Ryan, U.S., *Environ. Health Perspect.*, 56, 103, 1984. With permission.)

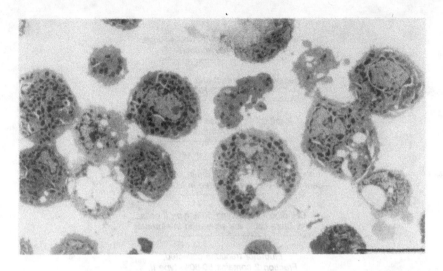

Figure 7a. Electron micrograph of freshly isolated nonciliated bronchiolar epithelial (Clara) cells showing a cuboidal morphology with eccentric nucleus, smooth endoplasmic reticulum and numerous electron dense secretory granules in the cytoplasm. Bar = 10 µm. (From Walker, S. et al., *Exper. Lung Res.*, 15, 553, 1989. With permission.)

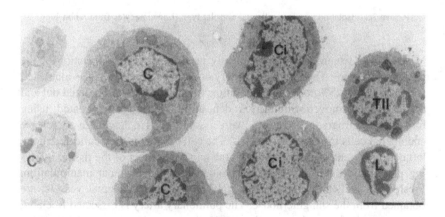

Figure 7b. Electron micrograph of several freshly isolated cells in the same culture demonstrating the ease of morphological identification. Nonciliated bronchiolar epithelial cells (C), ciliated cells (Ci), type II alveolar epithelial cells (TII), lymphocytes (L). Bar = 1 µm. (From Walker, S. et al., *Exper. Lung Res.*, 15, 553, 1989. With permission.)

of organ specific antigens on capillary endothelial cell membranes, as well as the unique profile of cell membrane glycoproteins, are examples of differences which exist in cells of various origins. Despite these structural and functional modifications, several techniques have been developed for the isolation of endothelial cells from lung.

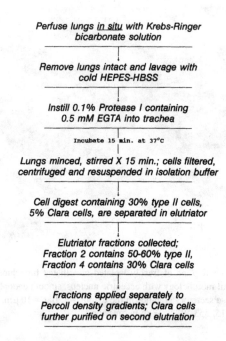

Figure 7c.   Isolation of Clara cells and alveolar type II pneumocytes from rabbit lung.

A method which has been frequently used is enzymatic digestion of the greater vessels.[33,55] The pulmonary artery or aorta of large animals is incubated with a 0.05 to 0.25% collagenase solution, after the vessel is dissected out and the fat layers removed. Similarly, rodent and rabbit lungs are treated, using forward or retrograde perfusion, with a solution of collagenase to release endothelial cells from smaller vessels. The cell effluent is then collected in the left coronary ventricle. Alternatively, the harsh treatment of the tissue involving proteolytic digestions may be avoided using mechanical manipulation. Fresh heart and lungs from young calves, with attached great vessels, are obtained from the slaughterhouse. The pulmonary artery is dissected free and, after removal of fat and connective tissue, the luminal surface of the vessel is scraped with a sterile scalpel. The cells are collected from the scalpel in a single sheet and transferred to a culture vessel, thus preserving the original polarity of the cell layer.

Microvessels are the primary sites of transvascular fluid and solute exchange and of endothelial injury in lungs. Del Vecchio et al.[58] describe the isolation of bovine endothelial cells from microvessels of minced lung using a collagenase solution. The cells are plated onto fibronectin-coated dishes and grown in RPMI 1640 containing 20% fetal bovine serum, heparin, and retinal-derived growth factor. Pulmonary microvessel endothelial cells are reported to have a unique profile of cell surface glycoproteins different from pulmonary artery, vein, and aortic cells.[57]

Cells isolated with proteolytic enzymes grow to form confluent monolayers. The cells are subcultured using 0.05% trypsin solution containing 0.02% EDTA in Ca^{2+}- and Mg^{2+}-free physiological salt solution, such as Puck's saline or Hank's balanced salt solution. The technique is similar to that described for routine passaging of continuous cell lines. In addition, endothelial cells are further purified using differential adhesion. That is, these cells are preferentially detached with trypsin and, in a mixture of cells, will settle before smooth muscle cells after inoculation of a culture flask. Thus, the selection of relatively pure populations can be achieved through this technique.

Primary cultures of pulmonary endothelial cells isolated mechanically are grown in flasks, and the monolayers are subcultured by scraping with a rubber policeman. Alternatively, the preferred method for handling these cells is by culturing on polyacrylamide or polystyrene microcarrier beads.[56] Endothelial cells attach to beads of all sizes, have the capacity for large yields in roller culture bottles, can be subcultured and transferred to flasks routinely, and are cloned without use of enzymes. The cells are also detached from the beads by gentle vortexing and filtering through large micron filters.

Endothelial cells are identified by morphological and biochemical criteria (Figures 5 and 6). In monolayer culture, they demonstrate a typical cobblestone-like arrangement at confluence, and they possess caveolae on the luminal surface, which is typical of these cells *in situ*. Functionally, the presence and measurable activity of angiotensin converting enzyme on the plasma membrane is a hallmark of the cells in culture. Several enzymes active in xenobiotic metabolism have also been detected in culture.[33,54]

Recently a spontaneously transformed endothelial cell line (ECV304) was described as having potential as a model for human endothelial cells.[59] Although the cell line, which is derived from human umbilical cord, has an infinite life span, it exhibits contact inhibition and demonstrates marked angiogenic activity. It expresses various endothelial cell adhesion molecules on the surface, reacts with an anti-endothelial cell autoantibody (present in several autoimmune diseases), and synthesizes several cell-specific bioactive substances, such as prostaglandin I_2 and thromboxane A_2.

5. Characterization of Lung Cells

Table 10 lists some of the features characteristic of the lung cell types which can be used to identify freshly isolated primary cultures. Some of these characteristics are described above and are also essential in monitoring for changes occurring during subculture.

TABLE 10
Structural Characteristics and Toxicological Parameters of Primary Cultures of Lung Cells

Cell type	Characteristic features	
	Structural	Toxicological
Clara cell	Non-ciliated cuboidal bronchiolar epithelia	Cytochrome P450 activity
Alveolar macrophages	Large, mononuclear cells of interstitium and epithelial surface of lung	Secretory activity and phagocytic properties
Type II alveolar cells	Irregular cuboidal epithelial cells containing numerous microvilli and lamellar bodies	Surfactant synthesis and secretion Cytochrome P450
Endothelial cells	Cobblestone appearance with caveolae on luminal surface	ACE[a] activity, xenobiotic metabolism

[a]   Angiotensin converting enzyme.

II.   GENERAL FUNCTIONAL MARKERS

Some specialized cells and the criteria used to assess their response to cytotoxicity are listed in Table 11. Because these systems can also be used to determine general toxicity, it is important to maintain the degree of differentiation of the cells at a level where functional characteristics are optimum. In this way organ-specific toxicity can be measured with greater certainty. Accurate measurement of the parameters, in turn, confirm that the cells have not dedifferentiated.

III.   DIFFERENTIAL TOXICITY

It is important to note that even validated and popular test systems may measure indirect effects of basal cytotoxicity, as was the case with the systems for mechanistic studies. Several studies have compared the sensitivity of organ-specific primary cultures to undifferentiated cell lines. With few exceptions, such as corneal cells from rabbit and kidney cells from monkey, the results have shown that the toxicity for most chemicals was similar for the different types of cells. This suggests that the tests using primary cultures of highly specialized cells for a majority of substances measure basal cytotoxicity, at least at the higher doses. It is possible that by studying specialized

TABLE 11
Primary Cultures of Specialized Cells and Their Functional Markers in Toxicity Testing

Cells	Organ	Functional Markers or Specialized Pathways
Hepatocytes	Liver	Leakage of specific enzymes (cytochrome P450); inhibition of biosynthesis and secretion of albumin, bile acids, urea
Macrophages	Lung	Inhibition of phagocytosis, chemotaxis, rosette formation
Epithelial cells	Lung	Phospholipid synthesis, inhibition of specific enzymes
Erythrocytes	Blood	Inhibition of glucose transport and glucose-6-phosphate dehydrogenase
Lymphocytes	Blood	Inhibition of lysolecithin and acyltransferase
Myelocytes	Bone marrow	Inhibition of colony formation
Mast cells	Lung	Inhibition of histamine production
Cardiac cells	Heart	Changes in spontaneous contractions, leakage of enzymes
Renal cells	Kidney	Inhibition of renin secretion, synthesis of metallothionein
Neurons	Brain, spinal cord	Inhibition of GABA-mediated postsynaptic transmission, changes in morphology
Glial cells	Brain	Glycosaminoglycan content, oxygen consumption
Muscle cells	Skeletal	Inhibition of acetylcholinesterase, morphology

functions of organ-specific cells, fine adjustments in the dosages used during exposure are required. Alternatively, toxicity may be demonstrated through a narrower range of dosages or shorter period of time. This would require that measurements be taken at intervals *during* the exposure period rather than at the end of the incubation. For example, by measuring the amount of newly synthesized proteins in lung cells using radiolabeled precursors, Barile et al.[60] have determined that some chemicals possess a steep dose-response curve. This indicates that the chemical has a narrow range of toxicity (Figures 8 and 9). Although protein synthesis is a fundamental process of cellular metabolism, the technique is sensitive enough to detect small changes in toxic response. Nevertheless, although basal cytotoxicity is probably being measured, it does not exclude the possibility that these cells are more sensitive to compounds with selective pulmonary toxicity.

Ekwall and Acosta[61] compared the cytotoxicity of 14 hepatotoxic substances in cell lines, including liver cells (Table 12). Most substances appear to induce liver injury *in vitro* through mechanisms involving basal cytotoxicity. This same explanation is valid for the toxicity observed in HeLa cells

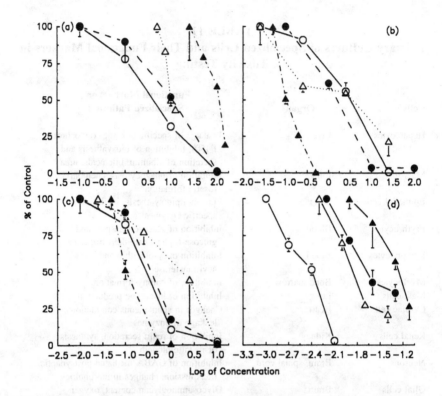

Figure 8. Concentration-response curves for representative chemicals using the [³H]proline assay comparing relative *in vitro* cytotoxicities. Inhibitory concentrations at the 50% level were determined from these plots. (a) Ethanol (○), methanol (●), isopropanol (△), ethylene glycol (▲); (b) acetylsalicylic acid (○), paracetamol (●), diazepam (△), propoxyphene (▲); (c) 2,4-dichlorophenoxyacetic acid (○), phenol (●), xylene (△), malathion (▲); (d) thioridazine (○), arsenic trioxide (△), mercuric chloride (●), propranolol (▲). (From Barile, F. A. et al., *Toxicol. In Vitro*, 7, 111, 1993. With permission.)

and Chang cells. These substances also cause relatively nonspecific liver injuries *in vivo*, such as biliary hepatitis and syndromes which mimic viral hepatitis. Four substances were more toxic in hepatocytes than other cell lines, i.e., they appeared to exhibit a higher **differential toxicity** for the hepatocytes. It was suggested that such information could be used to predict specific metabolite-induced liver injury in humans. Similar pathologies, such as centrilobular hepatocellular degeneration or necrosis, have been attributed to the untoward effects of metabolites of the parent compound. In addition, the results indicate that a large proportion of target organ toxicity is caused by the same mechanisms that are responsible for basal cytotoxicity in the organ. This suggests that primary cultures of hepatocytes can be used to screen for hepatotoxic agents, which is not possible using other differentiated cells.

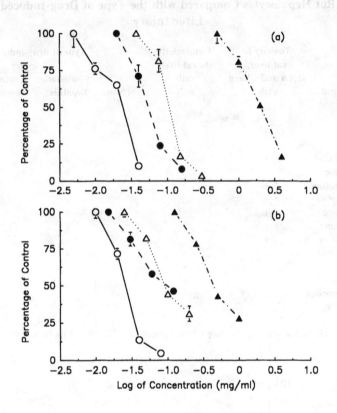

Figure 9. Comparisons of different chemicals and their concentration-effect curves from (a) 24-hour [³H]proline incorporation studies, and (b) 72-hour growth studies; pentachlorophenol (○), orphenadrine (●), sodium oxalate (△), and caffeine (▲). (Adapted from Hopkinson et al., *ATLA*, 21, 167, 1993.)

More recently, Hopkinson et al.[63] show that 24-hour protein synthesis measurement is as sensitive and reliable an indicator for detecting basal cytotoxicity as 72-hour growth studies using the same cell line. In addition, cytotoxicity did not differ between different cell lines in 72-hour growth studies (Figures 10 and 11). Thus, protein synthesis is a process which qualitatively does not differ from other indicators of cellular metabolism, such as cell proliferation or mitochondrial function. One must be aware, however, that as with any testing procedure, some chemicals may elude the cytotoxic screen, thus labeling some substances as having low toxic potential.

TABLE 12
Differential Toxicity of 14 Agents to Nondifferentiated Cell Lines and Rat Hepatocytes Compared with the Type of Drug-Induced Liver Injury

Compound	Toxicity to rat liver, HeLa and Chang cells	Cytotoxicity to rat liver cells only	None	Type of drug-induced liver injury	
				Cholestatic hepatitis	Centrilobular degeneration
Caffeine	+		+		
Ethanol	+			+	
Salicylic acid	+			+	
Papaverine	+			+	
Nitrofurantoin	+			+	
Amitriptyline	+			+	
Imipramine	+			+	
Chloriysime	+			+	
Desipramine	+			+	
Nortriptyline	+			+	
CCl$_4$		+			+
Halothane		+			+
Acetaminophen		+			+
Tetracycline		+			+

From Ekwall, B. and Acosta, D., *Drug Chem. Toxicol.*, 5, 229, 1987. With permission.

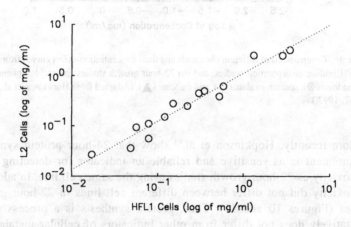

Figure 10. Comparison of IC$_{50}$s for MEIC chemicals numbers 31 to 50: 72-hour cell proliferation studies using continuous cell lines, rat lung epithelial cells (L2, immortal) vs. human fetal lung cells (HFL1, finite). For growth experiments, treated groups were incubated in Dulbecco's modified Eagle's medium supplemented with increasing concentrations of the chemicals 24 hours after initial seeding. On day 3, cells were trypsinized and counted using a haemocytometer (see Chapter 11, Section III, for description of MEIC program and Table 4 for list of chemicals).

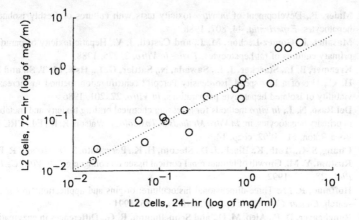

Figure 11. Comparison of IC_{50}s for MEIC chemicals numbers 31 to 50: 24- and 72-hour cytotoxicity studies using rat lung epithelial cells (L2, immortal). For 24-hour studies, monolayers were incubated with increasing concentrations of each chemical in at least 4 doses for 24 h at 37°C in an atmosphere of 10% CO_2/90% air in the presence of L-[2,3-³H]proline (1 uCi/ml). Labeling media consisted of modified Eagle's medium without glutamine (GIBCO's "Autoclavable MEM") supplemented with sodium bicarbonate, dialyzed fetal bovine serum (2%), and ascorbic acid (50 μg/ml). After 24 h, cells and media were harvested together and the TCA-insoluble material was counted. For 72-hour growth experiments, treated groups were incubated as described for Figure 10 (see Chapter 11, Section III, for description of MEIC program and Table 4 for list of chemicals). (Adapted from Hopkinson et al., *ATLA*, 21, 167, 1993.)

REFERENCES

1. **Gram, T. E., Okine, L. K., and Gram, R. A.**, The metabolism of xenobiotics by certain extrahepatic organs and its relation to toxicity, *Annu. Rev. Pharmacol. Toxicol.*, 26, 259, 1986.
2. **Bend, J. R. and Serabjit-Singh, C. J.**, Xenobiotic metabolism by extrahepatic tissues: relationship to target organ and cell toxicity, in *Drug Metabolism and Drug Toxicity*, Mitchell, J. R. and Horning, M. G., Eds., Raven Press, New York, 1984, 99–136.
3. **Boyd, M. R.**, Evidence for the Clara cell as a site of cytochrome P-450 dependent mixed function oxidase activity in lung, *Nature (London)*, 269, 713, 1977.
4. **Buckpitt, A. R., Statham, C. N., and Boyd, M. R.**, In vivo studies on the target tissue metabolism, covalent binding, glutathione depletion and toxicity of 4-ipomeanol in birds, species deficient in pulmonary enzymes for metabolic activation, *Toxicol. Appl. Pharmacol.*, 65, 38, 1982.
5. **Ecobichon, D. J.**, *The Basis of Toxicity Testing*, 1st ed., CRC Press, Boca Raton, FL, 1992, chap. 2.
6. **Gomez-Lechon, M. J., Montoya, A., Lopez, P., Donato, T., Larrauri, A., and Castell, J. V.**, The potential use of cultured hepatocytes in predicting the hepatotoxicity of xenobiotics, *Xenobiotica*, 18, 725, 1988.
7. **Castell, J. V., Montoya, A., Larrauri, A., Lopez, P., and Gomez-Lechon, M. J.**, Effects of benorylate and impacine on the metabolism of cultured hepatocytes, *Xenobiotica*, 15, 743, 1985.
8. **Jover, R., Ponsoda, X., Castell, J. V., and Gomez-Lechon, M. J.**, Evaluation of the cytotoxicity of ten chemicals on human cultured hepatocytes: predictability of human toxicity and comparison with rodent cell culture systems, *Toxic. In Vitro*, 6, 47, 1992.

9. **Maier, P.**, Development of *in vitro* toxicity tests with cultures of freshly isolated rat hepatocytes, *Experientia*, 44, 807, 1988.

10. **Masanet, J., Gomez-Lechon, M. J., and Castell, J. V.**, Hepatic toxicity of paraquat in primary cultures of rat hepatocytes, *Toxic. In Vitro*, 2, 275, 1988.

11. **Kreamer, B. L., Staecker, J. L., Sawada, N., Sattler, G. L., Hsia, M. T. S., and Pitot, H. C.**, Use of a low speed, iso-density Percoll® centrifugation method to increase the viability of isolated hepatocyte preparations, *In Vitro*, 22, 201, 1986.

12. **Del Raso, N. J.**, *In vitro* methods for assessing chemical or drug toxicity and metabolism in primary hepatocytes, in: *In Vitro Methods in Toxicology*, Watson, R. R., Ed., CRC Press, Boca Raton, FL, 1992, chap. 14.

13. **Chang, S.-G., Toth, K., Black, J. D., Slocum, H. K., Perrapato, S. D., Huben, R. P, and Rustum, Y. M.**, Growth of human renal cortical tissue on collagen gel, *In Vitro Cell. Dev. Biol.*, 28, 128, 1992.

14. **Hoffman, R. M.**, Three-dimensional histoculture: origins and applications in cancer research, *Cancer Cells (Cold Spring Harbor)*, 3, 86, 1991.

15. **Rodeheaver, D. P., Aleo, M. D., and Schnellmann, R. G.**, Differences in enzymatic and mechanical isolated rabbit renal proximal tubules; comparison in long-term incubation, *In Vitro Cell. Dev. Biol.*, 26, 898, 1990.

16. **Boogaard, P. J., Mulder, G. J., and Nagelkerke, J. F.**, Isolated proximal tubular cells from rat kidney as an *in vitro* model for studies on nephrotoxicity, *Toxicol. Appl. Pharmacol.*, 101, 144, 1989.

17. **Boogaard, P. J., Nagelkerke, J. F., and Mulder, G. J.**, Renal proximal tubular cells in suspension or in primary culture as *in vitro* models to study nephrotoxicity, *Chem. Biol. Interact.*, 76, 251, 1990.

18. **Handler, J. S.**, Studies of kidney cells in culture, *Kidney Int.*, 30, 208, 1986.

19. **Williams, P. D., Laska, D. A., Tay, L. K., and Hottendorf, G. H.**, Comparative toxicities of cephalosporin antibiotics in a rabbit kidney cell line (LLC-PK$_1$), *Antimicrob. Agents Chemother.*, 32, 314, 1988.

20. **Bacon, J. A., Linsemen, D. A., and Raczniak, T. J.**, *In Vitro* cytotoxicity of tetracyclines and aminoglycosides in LLC-PK$_1$, MDCK and Chang continuous cell lines, *Toxicol. In Vitro*, 4, 384, 1990.

21. **Ford, S. M., Laska, D. A., Hottendorf, G. H., and Williams, P. D.**, Assessment of the ability of several *in vitro* models to predict the nephrotoxicity of φ-lactam antibiotics, *Toxicologist*, 8, 187, 1988.

22. **Bach, P. H.**, Detection of chemically induced renal injury: the cascade of degenerative morphological and functional changes that follow the primary nephrotoxic insult and evaluation of these changes by *in vitro* methods, *Toxicol. Lett.*, 46, 237, 1989.

23. **Borenfreund, E. and Puerner, J. A.**, Toxicity determined *in vitro* by morphological alterations and neutral red absorption, *Toxicol. Lett.*, 24, 119, 1985.

24. **Sens, M. A., Hennigar, G. R., and Sens, D. A.**, An *in vitro* model for aminoglycoside nephrotoxicity: human proximal convoluted tubular cell culture, *Lab. Invest.*, 52, 61A, 1985.

25. **Kikkawa, Y. and Yoneda, K.**, The type II epithelial cell of the lung. I. Method of isolation, *Lab. Invest.*, 30, 76, 1974.

26. **Boyd, M., Statham, C., and Longo, N.**, The pulmonary Clara cell as a target for toxic chemicals requiring metabolic activation: studies with carbon tetrachloride, *J. Pharmacol. Exp. Ther.*, 212, 109, 1980.

27. **Devereux, J. R. and Fouts, J. R.**, A procedure for isolation of rabbit pulmonary epithelial cells for study of foreign compound metabolism, in *Microsomes, Drug Oxidations, and Chemical Carcinogenesis*, Cooh, M. J., Conney, A. H., Estabrook, R. W., Gelboin, H. V., Gilette, J. R., and O'Brien, P. J., Eds., Academic Press, New York, 1980, 825.

28. **Dobbs, L. G., Geppert, E. F., Williams, M. C., Greenleaf, R. D., and Mason, R. J.**, Metabolic properties and utlrastructure of alveolar type II cells isolated from elastase, *Biochim. Biophys. Acta*, 618, 510, 1980.

29. **Dobbs, L. G., Mason, R. J., Williams, M. C., Benson, B. J., and Sueishi, K.,** Secretion of surfactant by primary cultures of alveolar type II cells isolated from rats, *Biochim. Biophys. Acta,* 713, 118, 1982.

30. **Devereux, T. R., Serabjit-Singh, C. J., Slaughter, S. R., Wolf, C. R., Philpot, R. M., and Fouts, J. R.,** Identification of cytochrome P-450 isozymes in nonciliated bronchiolar epithelial (Clara) and alveolar type II cells isolated from rabbit lung, *Exp. Lung Res.,* 2, 221, 1981.

31. **Devereux, T. R. and Fouts, J. R.,** Isolation and identification of Clara cells from rabbit lung, *In Vitro,* 16, 958, 1980.

32. **Devereux, T. R. and Fouts, J. R.,** Isolation of pulmonary cells and use in studies of xenobiotic metabolism, in *Methods in Enzymology,* Vol. 71, Jacoby, J., Ed., Academic Press, New York, 1981, 147–154.

33. **Ryan, U. S.,** Isolation and culture of pulmonary endothelial cells, *Environ. Health Perspect.,* 56, 103, 1984.

34. **Devereux, T. R.,** Alveolar type II and Clara cells: isolation and xenobiotic metabolism, *Environ. Health Perspect.,* 56, 95, 1984.

35. **Brown, S. E. S., Goodman, B. E., and Crandall, E. D.,** Type II alveolar epithelial cells in suspension: separation by density and velocity, *Lung,* 162, 271, 1984.

36. **Bend, J. R., Serabjit-Singh, C. J., and Philpot, R. M.,** Pulmonary uptake, accumulation and metabolism of chemicals, *Annu. Rev. Pharmacol. Toxicol.,* 25, 97, 1985.

37. **Loo, C. K. C., Smith, G. J., and Lykke, A. W. J.,** Effects of hyperoxia on surfactant morphology and cell viability in organotypic cultures of fetal rat lung, *Exp. Lung Res.,* 15, 597, 1989.

38. **Ryrfeldt, A., Cotgreave, I. A., and Moldeus, P.,** *In vitro* models to study mechanisms of lung toxicity, *ATLA,* 18, 267, 1990.

39. **Banks, M. A., Martin, W. G., Pailes, W. H., and Castranova, V.,** Taurine uptake by isolated alveolar macrophages and type II cells, *J. Appl. Physiol.,* 66, 1079, 1989.

40. **Barile, F. A., Arjun, S., and Senechal, J.-J.,** Paraquat alters growth, DNA and protein synthesis in lung epithelial cell and fibroblast cultures, *ATLA,* 20, 251, 1992.

41. **Douglas, W. H. J., Moorman, G. W., and Teel, R. W.,** Formation of histopathic structures from monodisperse fetal lung cells cultured on a three-dimensional substrate, *In Vitro,* 12, 373, 1976.

42. **Cutz, E. and Conen, P. E.,** Ultrastructure and cytochemistry of Clara cells, *Am. J. Pathol.,* 62, 127, 1971.

43. **Reasor, M. J., Nadeau, D., and Hook, G. E. R.,** Extracellular alkaline phosphatase in the airways of the rabbit lung, *Lung,* 155, 321, 1978.

44. **Devereux, T. R., Jones, K. G., Bend, J. R., Fouts, J. R., Boyd, M. R., and Statham, C. N.,** *In vitro* metabolic activation of the pulmonary toxin, 4-ipomeanol, in nonciliated bronchiolar epithelial (Clara) and alveolar type II cells isolated from rabbit lung, *J. Pharmacol. Exp. Therap.,* 220, 223, 1981.

45. **Kehrer, J. R. and Witschi, H.,** Effects of drug metabolism inhibitors on butylated hydroxytoluene-induced pulmonary toxicity in mice, *Toxicol. Appl. Pharmacol.,* 53, 333, 1980.

46. **Boyd, M., Statham, C. N., Franklin, R., and Mitchell, J.,** Pulmonary bronchiolar alkylation and necrosis by 3-methylfuran, a naturally-occuring potential atmospheric pollutant, *Nature,* 272, 270, 1978.

47. **Haugen, A. and Aune, T.,** Culture of rabbit pulmonary Clara cells, *Proc. Soc. Exp. Biol. Med.,* 182, 277, 1986.

48. **Massaro, G. D., Fischman, C., Chiang M.-J., Amado, C., and Massaro, D.,** Regulation of secretion in Clara cells studied using the isolated perfused rat lung, *J. Clin. Invest.,* 67, 345, 1981.

49. **Walker, S. R., Hale, S., Malkinson, A. M., and Mason, R. J.,** Properties of isolated nonciliated bronchiolar cells from mouse lung, *Exp. Lung Res.,* 15, 553, 1989.

50. **Walker, S. R., Williams, M. C., and Benson, B.**, Immunocytochemical localization of the major surfactant apoproteins in type II cells, Clara cells and alveolar macrophages of rat lung, *J. Histochem. Cytochem.*, 34, 1137, 1986.

51. **Bend, J. R., Horton, J. K., Brigelius, R., Dostal, L. A., Mason, R. P., and Serabjit-Singh, C. J.**, Cell selective toxicity in the lung: role of metabolism, in *Basic Science in Toxicology*, Volans, G. N., Sims, J., Sullivan, F. M., and Turner, P., Eds., Proc. of the 5th Int. Congr. of Toxicology, Brighton, U.K., Taylor and Francis, London, 1990, 220.

52. **Boyd, M. R., Burka, L. T., Wilson, B. J., and Sasame, H. A.**, *In vitro* studies on the metabolic activation of the pulmonary toxin 4-ipomeanol by rat lung and liver microsomes, *J. Pharmacol. Exp. Ther.*, 207, 677, 1978.

53. **Ryan, U. S., Ryan, J. W., Whitaker, C., and Chiu, A.**, Localization of angiotensin converting enzyme (Kinase II): immunocytochemistry and immunoflourescence, *Tissue Cell*, 8, 125, 1976.

54. **Ryan, U. S., Clements, E., Habliston, D., and Ryan, J. W.**, Isolation and culture of pulmonary artery endothelial cells, *Tissue Cell*, 10, 535, 1978.

55. **Habliston, D., Whitaker, C., Hart, M. A., Ryan, U. S., and Ryan, J. W.**, Isolation and culture of endothelial cells from the lungs of small animals, *Am. Rev. Respir. Dis.*, 119, 853, 1979.

56. **Ryan, U. S., Mortara, M., and Whitaker, C.**, Methods for microcarrier culture of bovine pulmonary artery endothelial cells avoiding the use of enzymes, *Tissue Cell*, 12, 619, 1980.

57. **Lipton, B. H., Bensch, K. G., and Karasek, M. A.**, Microvessel endothelial cell transdifferentiation: phenotypic characterization, *Differentiation*, 46, 117, 1991.

58. **Del Vecchio, P. J., Siflinger-Birnboim, A., Belloni, P. N., Holleran, L. A., Lum, H., and Malik, A. B.**, Culture and characterization of pulmonary microvascular endothelial cells, *In Vitro Cell. Dev. Biol.*, 28A, 711, 1992.

59. **Heffelfinger, S. C., Hawkins, H. H., Barrish, J., Taylor, L., and Darlington, G. J.**, SK HEP-1: A human cell line of endothelial origin, *In Vitro Cell. Dev. Biol.*, 28, 136, 1992.

60. **Barile, F. A., Guzowski, D. E., Ripley, C., Siddiqi, Z., and Bienkowski, R. S.**, Ammonium chloride inhibits basal degradation of newly synthesized collagen in human fetal lung fibroblasts, *Arch. Biochem. Biophys.*, 276, 125, 1990.

61. **Ekwall, B. and Acosta, D.**, *In vitro* comparative toxicity of selected drugs and chemicals in HeLa cells, Chang liver cells, and rat hepatocytes, *Drug Chem. Toxicol.*, 5, 229, 1987.

62. **Barile, F. A., Arjun, S., and Hopkinson, D.**, *In vitro* cytotoxicity testing: biological and statistical significance, *Toxic. In Vitro*, 7, 111, 1993.

63. **Hopkinson, D., Bourne, R., and Barile, F. A.**, *In vitro* cytotoxicity testing: 24- and 72-hour studies with cultured lung cells, *ATLA*, 21, 167, 1993.

64. **Houser, S., Borges, L., Guzowski, D. E., and Barile, F. A.**, Isolation and maintenance of continuous cultures of epithelial cells from chemically-injured adult rabbit lung, *ATLA*, 17, 301, 1990.

Chapter 5

CELLULAR METHODS OF LOCAL TOXICITY

I. *IN VITRO* CELL MODELING SYSTEMS

Historically, testing of chemicals for their ability to cause local toxicity has been the subject of cell modeling systems. This testing has been performed in skin explants and isolated tissues, as well as in cultured cells. Until now, the results obtained with these models have not been successful in predicting toxicity encountered *in vivo* principally because of the pharmacokinetics of *in vivo* administration — that is, rapid elimination from the site of administration *in vivo* results in artificially lower toxic concentrations. Recently, technical advances and improved knowledge of cell culture technology has allowed for improvements in the development of cellular systems.

The development of a cellular model of local toxicity involves the establishment of primary cultures from specialized tissues, including dermal, lung, ocular, and dental epithelium. These primary cultures possess specific functional markers which can distinguish selective toxic effects. For example, the effect of a chemical on a specific functional target, which is unique for ocular epithelia, can be assessed and may help to distinguish between the chemical's effect on ocular cells and any other cell type. Other isolated *in vitro* models used to screen for local toxicity incorporate several tissue types. These include systems which have been developed to detect general eye irritancy and toxicity of implants. The tissue types are not limited to the organ of origin but include supportive structures, such as connective tissue cells. Recently, corneal epithelial cell culture models have been introduced to test for chemicals with potential ocular effects. However, dedifferentiated continuous cell lines are still being used to screen compounds for toxicity, which run the risk of measuring basal cytotoxic phenomena.

II. TESTING OF DENTAL MATERIALS

The testing of materials used in the manufacturing of surgical prostheses and their plastic components have allowed for the establishment of cellular models. The relevance of these systems has been applied and appreciated through extensive practical use. Original applications of testing for local irritancy included the use of tissue cultures for screening dental materials, while a few standard methods incorporating cell lines have been established internationally. Examples of these methods include (1) The agar overlay test;[1,2,18] (2) Wennberg's millipore filter method;[3] and (3) Spangberg's chromium-leakage method.[4]

The first two methods are designed to test solid materials. The material to be analyzed is placed on an agar overlay or on a Millipore filter and the solubilized material diffuses through the porous semipermeable barriers until it encounters the cells. An endpoint determination is then measured to determine the degree of toxicity. Results of testing some dental alloys, separately and in combinations, showed poor correlation between the agarose overlay tests, the hemolysis tests, and the implantation tests in rodents.[5] Only the *in vivo* implantation test and the *in vitro* toxicity test using macrophages showed a fairly good correlation.

Spangberg's chromium leakage method[4] uses L929 mouse fibroblasts which are labeled overnight with ^{51}Cr. The material tested serves as the substrate for the growth of the cells and thus completely covers the bottom of the testing wells. The target cells are then layered over the material, thus establishing close contact during the entire test period. Release and measurement of ^{51}Cr during a 4- or 24-hour incubation period is proportional to the toxicity of the material.

Recently, the toxicity of dental materials has been evaluated using a three-level approach:

1. Cytotoxicity determined *in vitro* using cell culture techniques
2. Animal studies using implanted materials in subcutaneous tissue or muscle
3. Evaluating the pulp reactions of 'usage tests' after insertion of restorations in animal or human teeth

Results of the testing of chemicals from many of these tests reveal poor correlations to *in vivo* data.[8] Apparently *in vitro* cell culture techniques measure cytotoxicity and, therefore, may not reflect the situations commonly encountered in clinical practice. In addition, several technical aspects of the cytotoxicity tests vary significantly among the different tests, as well as with *in vivo* experience. For instance, the time of incubation with the test chemical, usually 24 hours, represents a short period of contact with gingival tissue which does not mimic the years of exposure to dental materials. The toxic concentrations of the test substance in culture does not correspond to concentrations detected *in vivo* in the target tissue. This is apparently a result of the inability of the substance to penetrate intact skin or because the chemical is rapidly diluted from the target tissue through the action of blood flow. Consequently, the concentration of the chemical at the active site is not analogous to local pharmacokinetics exhibited *in vivo*. Thus, the data suggest that more than one *in vitro* test should be performed or the tests should be refined to meet specific dental requirements.[5-8]

III.   TESTING OF OCULAR IRRITANTS

Table 1 summarizes some of the important studies which dictated the direction of the development of cytotoxicity tests for ocular irritancy. Most of the tests have been successful in obtaining good correlations with the use of continuous cell lines. These tests have used monolayer cultures in conjunction

TABLE 1
Summary of Recent Ocular Toxicity
Testing Studies

Test cells	Assay	Test chemicals	Correlation to in vivo data[a]
Hep G2 BALB/c 3T3 V79, murine macrophages Rabbit corneal epithelial cells	7-day colony formation after 24-h incubation	34 chemicals	Good agreement among 5 cell lines[46]
Rabbit corneal cells	7-day colony formation after 24 h incubation	Surfactants, shampoos	Correlation coef- ficient = 0.9[29]
Baby hamster kidney fibroblasts	4 h cell detachment 48 h growth inhibi- tion, 7 day cloning efficiency	57 various organic & inorganic chemicals	80% predictive of in vitro classification[11]
BALB/c 3T3	24 h neutral red (NR) uptake	Anionic surfactants 35 different chemicals	NR uptake in agreement[47]
BALB/c 3T3	24 h protein 4 h uridine uptake	50 chemicals	Rank correlation in agreement[10,12]
Enucleated rabbit eyes	Corneal thickness up to 5 min exposure	11 chemicals	Broad correlation[49]
Primary rabbit corneal epithelial cells	[³H]thymidine incorporation	Antineoplastic agents	Correlation with in vivo ocular irritation[25,26]
Embryonated chicken eggs	HET-CAM assay	60 chemicals & 41 cosmetic formulations	High rank correlation with eye irritants in vivo[16]
Isolated rabbit cornea	Changes in elec- trical impedance	Anionic surfactants	Correlation with in vivo ocular irritation[48]

[a]   Correlation with available Draize eye test data; references are noted as superscripts.

with a variety of quantifiable cytotoxic indicators including neutral red uptake,[9] uridine uptake,[10] growth inhibition,[11] and chemotaxis.[12] The assay systems have shown usefulness in predicting the rank ordering of the ocular irritation potential of a number of test substances. Recently, Roguet et al.[13] studied 30 surfactants using a cell line derived from rabbit cornea (SIRC) with the neutral red uptake bioassay as the indicator system. The results reveal that a good correlation ($r = 0.80$) was obtained, while false negatives were minimal (6%). Thus, it appears that continuous cell lines are suitable as indicators of eye

TABLE 2
Comparison of Ocular Irritancy and Cytotoxicity of 18 Compounds Measured in Three Different Cell Systems

Compound	Ocular irritancy[a]	MIT-24[b] (mM)	UI-50[c] (mM)	HTD[d] (mM)
1. Benzalkonium chloride	Severe	0.0008	0.002	0.003
2. Sodium lauryl sulfate	Severe-mod	0.11	0.31	0.28
3. Sodium hypochlorite	Severe	0.54	1.0	1.06
4. Trichloroacetic acid	Severe-mod	4.5	24.	12.2
5. 1-Octanol	Severe	9.5	0.9	6.4
6. Solketal	Moderate	14.	83.	40.0
7. 1-Pentanol	Severe-mod	16.	14.	12.9
8. 1-Heptanol	Severe	42.	2.8	8.2
9. 1-Butanol	Moderate	68.	77.	65.4
10. Isopropyl alcohol	Mild-mod	80.	268.	104.0
11. Tetrahydrofurfuryl alcohol	Moderate	110.	103.	61.8
12. Allyl alcohol	Severe	130.	12.	14.7
13. N-Methylformamide	Mild-mod	230.	411.	307.8
14. Ethylene glycol	Mild-mod	250.	456.	177.4
15. Dimethyl sulfoxide	Mild	280.	451.	310.2
16. Ethanol	Mild-mod	300.	541.	428.2
17. Propylene glycol	Mild	420.	476.	435.2
18. Methanol	Mild-mod	650.	668.	864.5

[a]   Results from Draize tests of irritancy in rabbits;[10]

[b]   The minimal concentration of substance inhibiting the spreading from round to spindle-shaped form of approximately 50% of HeLa cells in cultures; calculated as the geometrical mean between the 24 h total inhibitory concentration and the highest concentration not inhibiting spreading;

[c]   The concentration of a substance inducing a 50% decrease in [³H]uridine uptake in Balb/c 3T3 cells after four hours of exposure *in vitro*, calculated by linear regression from dose-response curves[10];

[d]   The highest tolerated dose which induces only a slight degree of any morphological change in Balb/c 3T3 cells, determined microscopically at 24 h incubation time.[46]

From Selling, J. and Ekwall, B., *Xenobiotica*, 15, 511, 1985. With permission.

irritancy, implying that the basal cytotoxic action of most of these toxicants plays a critical role in screening for ocular toxicity.

Table 2 compares the results obtained from screening ocular irritants with three different *in vitro* tests.[14] The data for determining the potential ocular toxicity of these chemicals and solvents have been traditionally evaluated in rabbits using the Draize method. The results indicate that good correlations can be obtained among different animal and cellular systems. In addition, the different cellular systems uniformly ranked toxicity according to the toxic concentrations.

Besides being criticized for the use of the rabbit eye as a test system, comparison of the results from the Draize eye test[15] among investigators has contributed to large interlaboratory variability. As a consequence, the overwhelming pressures in the scientific and regulatory communities to find alternatives to the traditional Draize procedure have prompted the development of new or modified *in vitro* assays for monitoring ocular irritancy.

Using a modification of the hen's egg test-chorioallantoic membrane (HET-CAM) test, de Silva et al.[16] show a high rank order correlation between the scores given from testing 60 chemicals and 41 cosmetic formulations with Draize *in vivo* data. The modified procedure reduced the number of false positives traditionally associated with the CAM test.[17] The method yields information on the changes that occur in blood vessels as a response to local injury, rather than monitoring direct cytotoxicity. Combrier and Castelli[2] have modified the original protocol for the *in vitro* agarose overlay method[18] to include several classes of agarose classification, such as non-irritant, minimally irritant, mildly irritant, and irritant. The results of testing 56 different cosmetic formulations were correlated with *in vivo* Draize data. High correlations were obtained between the agarose overlay method and the Draize test (86% concordance value). Boue-Grabot et al.[19] used human fibroblasts cultured on microporous membranes to determine the cytotoxicity of non-hydrosoluble substances. The monolayer rests on the surface of the membrane; the latter acts as an interface between the monolayer above and the test substance dissolved in a non-aqueous, non-cytotoxic medium below. The results of testing 169 cosmetic and pharmaceutical products revealed that the *in vitro* test had a sensitivity, specificity, and predictivity of 86%, 89%, and 67%, respectively, when compared to the Draize procedure.

The EYTEX™ system was designed as a biochemical procedure to evaluate ocular irritation.[20] The system is based on the concept that the normal transparent state of the cornea depends on the relative degree of hydration and organization of corneal proteins. Corneal opacification, therefore, is a result of a decrease in protein hydration associated with changes in protein conformation and aggregation. The EYTEX™ test simulates corneal opacification by using alterations in the hydration and conformation of an ordered macromolecular matrix to predict *in vivo* ocular irritancy. In a study conducted to screen for ocular irritancy, 465 cosmetic products and raw formulations were evaluated. When compared with historical *in vivo* data, the EYTEX™ system showed a positive agreement, with a concordance value of 80% and an 85% predictability.[20] An assay error of 20% was due primarily to overestimation of some non-irritant formulations. Courtellemont et al.[21] independently confirmed these results with the EYTEX™ system after testing 52 formulations and 49 chemicals. They suggest that the test could be a reliable *in vitro* prescreening system for determining the eye irritation potential of cosmetic formulations.

Regnier and Imbert[22] identified 74% of severely irritant, and 97% of non-severely irritant chemicals, of 166 substances tested according to pH and acid/alkaline reserve. They suggest that analysis of physicochemical parameters represents an important first line screen for ocular irritation.

The use of cell culture methods for assessing eye toxicity has been successful as a result of the similarity in the test methods.[46–49] In fact, the cell culture models incorporate several features which are consistent with the animal testing methods:

1. The incubation times *in vitro* are comparable to those used *in vivo*.
2. The test substance comes in direct contact with the eye epithelium as it does with the isolated cells.
3. The lack of blood vessels in the cornea allows for the attainment of steady state drug concentrations in contact with the tissue; thus the results obtained with the test protocols are not skewed by interruptions in absorption or local pharmacokinetic phenomena.
4. The toxicity may be due, in part, to basal cytotoxicity, which permits the use of continuous cell lines.

Alternatively, caution must be exercised with certain aspects of cell culture methodology when screening for ocular irritancy. For example, false positive and false negative results are usually errors of estimation and are closely related to the physicochemical properties of the test substance. Severely irritating properties of a substance *in vivo*, for instance, may be accounted for by extreme pH, which may not be detected *in vitro*. Because the cell culture methods described above are very sensitive, however, many non-irritating chemicals may be positively identified. In addition, several technical problems may be encountered, such as difficulty with solubility of the test substance in the culture medium, which may underestimate true toxic concentrations.[23-29]

Consequently, with the development of present methods and the introduction of new systems, it appears that it will be possible to define a battery of tests to replace the Draize procedure. In fact, the Cosmetic, Toiletry and Fragrance Association (CTFA) Evaluation of Alternatives Program[30] is currently assessing the effectiveness and limitations of *in vitro* tests in predicting the ocular irritation potential of generic cosmetic product formulations. The purpose of the program is to analyze and correlate the relationship between Draize primary eye irritation test data and comparative data from a selection of *in vitro* methods. The program is divided into three phases designed to test hydroalcoholic personal-care formulations (phase I), oil-water emulsion formulations (phase II), and surfactant based personal-care substances (phase III).[30,50] Several protocols have been evaluated in all three phases including: the agarose diffusion method, chorioallantoic membrane (CAM) assays, dual dye staining procedure, EYTEX™ system, neutral red release and uptake assays, pollen tube test, SIRC cytotoxicity test, tetrahymena motility assay, and total cellular protein procedure. Although the program does not claim to be validating the procedures, its intention is to provide the cosmetic industry with information on the performance of the usefulness of these protocols.

Recently, an international validation program was designed by COLIPA, the European cosmetic industry trade association, to assess *in vitro* eye irritation test performance.[50] Assays previously shown to be promising alternatives in the CTFA program are being evaluated in the COLIPA program for interlaboratory variation. Validation efforts are coordinated with the Japanese Cosmetic Industry Trade Association (JCITA) in consultation with CTFA.

IV. TESTING OF LOCAL IRRITANTS

Screening techniques to detect local toxicity have been used to investigate the effects of various irritating substances in the lung and on the skin. Agents which have been thoroughly studied in lung cell cultures include tobacco smoke,[31] diesel exhausts,[32] nitrogen dioxide,[33] aldehydes,[34] phorbol esters,[35] and paraquat.[36,37] The isolation of alveolar epithelial cells and non-ciliated bronchiolar epithelial (Clara) cells, and their establishment in primary culture, has greatly accelerated the underlying mechanisms of the local pulmonary response to xenobiotics. The culturing conditions, however, are not without problems and are often labor-intensive. In addition, continuous cultures of these cells are difficult to maintain in their differentiated state. Despite this, the need for such isolation methods is clear and is especially necessary in the case of tracheobronchial and bronchial epithelial cell types, particularly since they are known to be sensitive to a variety of inhaled xenobiotics. Details of the isolation and culture techniques, and cytotoxicity studies of lung cells, are outlined in Chapter 4 (Section I.C).

Earlier attempts to assess skin irritation of chemicals *in vitro* have met with considerable setbacks. The deficiencies associated with dermal fibroblasts, as a model for skin toxicity studies, center around the continuing controversy of the dedifferentiation of the cells once they are maintained in serial culture.[38] As a result they lose their phenotypic appearance and characteristic features of the original parent cell *in vivo*. Recent progress in toxicological studies has concentrated on the development of human epidermal keratinocytes, in contrast to previously used dermal fibroblasts.

Rheinwald and Green[39] markedly advanced human epidermal keratinocyte culture using mitotically inhibited "feeder layers" of 3T3 fibroblasts. Improvements on the feeder layer method involved the addition of epidermal growth factor,[40] cholera toxin,[41] or the use of culture dishes coated with extracellular matrix components.[42] In addition, the recent development of serum-free, hormone-supplemented media has enhanced the maintenance of keratinocyte cultures.[43] Normal human keratinocytes serially cultivated in defined medium maintained sustained growth and the ability to develop normally into a morphologically differentiated epidermis.

In general, cytotoxicity studies using cultures of keratinocytes have shown conflicting results when correlations are attempted between *in vitro* data and animal skin irritation tests for a large series of chemicals.[38] For skin irritation studies, the effective concentration of a chemical applied to the skin and reaching cells *in vivo* is a function of the pharmacokinetics of the chemical and the epidermal barrier. Consequently, skin cell culture model systems using

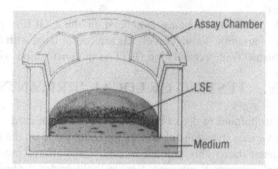

Figure 1.   Diagram of a cell culture model of TestSKIN™ (Organogenesis Inc., Cambridge, MA) showing the living skin equivalent (LSE) within its assay chamber. When the chamber is placed within the assay tray, the dermal portion of the LSE is in contact with the culture medium, while the epidermal portion is exposed to the air. This product has since been removed from the market by the company.

monolayer cultures may not be appropriate for screening chemicals with potential toxicity for skin. Interestingly, Wallace et al.[44] used the neutral red uptake assay to show that human epidermal keratinocytes, cultured in serum-free medium, were useful in predicting *generalized* acute lethal toxicity.

Recently, organotypic skin models have been developed which mimic human skin. TestSKIN™ LSE/LDE™, Matrex™ and Genesis™ (Organogenesis Inc., Cambridge, MA) were introduced as examples of cell culture model systems useful for toxicity testing of dermatologic products, as well as a wide range of chemicals, radiation, and pathogens. The test kits contained human skin fibroblasts embedded in three-dimensional living tissue that mimicked critical morphological and bio-chemical properties of actual skin (Figure 1). The models have been used for examining toxic responses to a variety of inflammatory agents, as well as for their ability to screen potentially irritating dermal and ocular agents.[45] These products, however, have since been removed from the market by the company.

Other products recently introduced, such as EpiDerm® (MatTek Corpora-tion, Ashland, MA) have improved on previous models. This product mimics human epidermis and is based on neonatal, foreskin-derived normal human epidermal keratinocytes in serum-free cell culture inserts. Analysis of the stratum corneum using transmission electron microscopy reveals a pattern characteristic of intercellular lamellae of native human epidermis. Unlike the previous models, EpiDerm® appears to have hemidesmosomes arranged in a well-organized basement membrane.

REFERENCES

1. **Autian, J.**, The new field of plastics toxicology: methods and results, in *CRC Critical Reviews in Toxicology*, Vol. 2, Goldberg, L., Ed., CRC Press, Cleveland, 1973, 1–40.
2. **Combrier, E. and Castelli, D.**, The agarose overlay method as a screening approach for ocular irritancy: application to cosmetic products, *ATLA*, 20, 438, 1992.

3. **Wennberg, A., Hasselgren, G., and Tronstad, L.**, A method for toxicity screening of biomaterials using cells cultured on millipore filters, *J. Biomed. Mater. Res.*, 13, 109, 1979.
4. **Pascon, E. A. and Spangberg, L. S. W.**, *In vitro* cytotoxicity of root canal filling materials. 1. Gutta-percha, *J. Endodontics*, 16, 429, 1990.
5. **Syrjanen, S., Hensten-Pettersen, A., Kangasniemi, K., and Yli-Urpo, A.**, *In vitro* and *in vivo* biological responses to some dental alloys tested separately and in combinations, *Biomaterials*, 6, 169, 1985.
6. **Mjor, I. A.**, Current views on biological testing of restorative materials, *J. Oral Rehab.*, 17, 503, 1990.
7. **Meryon, S. D. and Brook, A. M.**, *In vitro* cytotoxicity of three dentine bonding agents, *J. Dent.*, 17, 279, 1989.
8. **Arenholt-Bindslev, D., Horsted-Bindslev, P., and Philipsen, H. P.**, Toxic effects of two dental materials on human buccal epithelium *in vitro* and monkey buccal mucosa *in vitro*, *Scand. J. Dent. Res.*, 95, 467, 1987.
9. **Borenfreund, E. and Shopsis, C.**, Toxicity monitored with a correlated set of cell culture assays, *Xenobiotica*, 15, 705, 1985.
10. **Shopsis, C. and Sathe, S.**, Uridine uptake as a cytotoxicity test: correlations with the Draize test, *Toxicology*, 29, 195, 1984.
11. **Reinhardt, C. A., Pelli, D. A., and Zbinden, G.**, Interpretation of cell toxicity data for the estimation of potential irritation, *Food Chem. Toxicol.*, 23, 247, 1985.
12. **Shopsis, C., Borenfreund, E., Walberg, J., and Stark, D. M.**, A battery of potential alternatives to the Draize test: uridine uptake inhibition, morphological cytotoxicity, macrophage chemotaxis and exfoliate cytology, *Food Chem. Toxicol.*, 23, 259, 1985
13. **Roguet, R., Dossou, K, G., and Rougier, A.**, Prediction of eye irritation potential of surfactants using the SiRC-NRU cytotoxicity test, *ATLA*, 20, 451, 1992.
14. **Selling, J. and Ekwall, B.**, Screening of eye irritancy using HeLa cells, *Xenobiotica*, 15, 511, 1985.
15. **Draize, J. H., Woodward, G., and Clavery, H. O.**, Methods for the study of irritation and toxicity of substances applied to skin and mucous membrane, *J. Pharmacol. Exp. Ther.*, 82, 377, 1944.
16. **DeSilva, O., Rougier, A., and Dossou, K. G.**, The HET-CAM test: a study of the irritation potential of chemicals and formulations, *ATLA*, 20, 432, 1992.
17. **Leupke, N. P. and Kemper, F. H.**, The HET/CAM test: an alternative to the Draize eye test, *Food Chem. Toxicol.*, 24, 495, 1986.
18. **Rosenbluth, S. A., Weddington, G. R., Guess, L. W, and Autian, J.**, Tissue culture method for screening toxicity of plastics to be used in medical practice, *J. Pharm. Soc.*, 54, 156, 1965.
19. **Boue-Grabot, M., Halaviat, B., and Pinon, J.-F.**, Cytotoxicity of non-hydrosoluble substances towards human skin fibroblasts cultured on microporous membrane: a model for the study of ocular irritancy potential, *ATLA*, 20, 445, 1992.
20. **Kruszewski, F. H., Hearn, L. H., Smith, K. T., Teal, J. T., Gordon, V. C., and Dickens, M. S.**, Application of the EYTEX™ system to the evaluation of cosmetic products and their ingredients, *ATLA*, 20, 146, 1992.
21. **Courtellemont, P., Hebert, P., and Redziniak, G.**, Evaluation of the EYTEX™ system as a screening method for tolerance: application to raw material and finished products, *ATLA*, 20, 466, 1992.
22. **Regnier, J.-F. and Imbert, C.**, Contributions of physicochemical properties to the evaluation of ocular irritation, *ATLA*, 20, 457, 1992.
23. **Frazier, J. M.**, Update: a critical evaluation of alternatives to acute ocular irritancy testing, in *Alternative Methods in Toxicology*, Vol. 6, Goldberg, A. M., Ed., Mary Ann Liebert, New York, 1988, 67–75.
24. **Nardone, R. M. and Bradlaw, J. A.**, Toxicity testing with *in vitro* systems. I. Ocular tissue culture, *J. Toxicol. Cutaneous Ocular. Toxicol.*, 2, 81, 1983.
25. **Botti, R. E., Lazarus, H. M., Imperia P. S., and Lass J. H.**, Cytotoxicity of antineoplastic agents on rabbit corneal epithelium, *J. Toxicol. Cutaneous Ocular Toxicol.*, 6, 79, 1987.

26. **Lazarus, H. M., Imperia, P. S., Botti, R., Mack, R. J., and Lass, J. S.**, An *in vitro* method which assesses corneal epithelial toxicity due to antineoplastic, preservative, and antimicrobial agents, *Lens Eye Toxic. Res.*, 6, 59, 1989.
27. **Guillot, R.**, Ocular irritation: present cell culture models and perspectives, *ATLA*, 20, 471, 1992.
28. **Wilcox, D. K. and Bruner, L. H.**, *In vitro* alternatives for ocular safety testing: an outline of assays and possible future developments, *ATLA*, 18, 117, 1990.
29. **North-Root, H., Yackovich, F., Demetrulias, J., Gacula, M., and Heinze, J. E.**, Evaluation of an *in vitro* cell toxicity test using rabbit corneal cells to predict the eye irritation potential of surfactants, *Toxicol. Lett.*, 14, 207, 1982.
30. **Gettings, S. D., Bagley, D. M., Chudkowsky, M., Demetrulias, J. L., Dipasquale, L. C., Galli, C. L., Gay, R., Hintze, K. L., Janus, J., Marenus, K. D., Muscatiello, M. J., Pape, W. J. W., Renskers, K. J., Roddy, M. T., and Schnetzinger, R.**, Development of potential alternatives to the Draize test: the CTFA Evaluation of Alternatives Program. Phase II: review of materials and methods, *ATLA*, 20, 164, 1992.
31. **Ryrfeldt, A., Kroll, F., Berggren, M., and Moldeus, P.**, Hydroperoxide and cigarette smoke induced effects on lung mechanics and glutathione status in rat isolated perfused and ventilated lungs, *Life Sci.*, 42, 1439, 1988.
32. **Zamora, P. O., Gregory, R. E., Li, A. P., and Brooks, A. L.**, An *in vitro* model for the exposure of lung alveolar epithelial cells to toxic gases, *J. Environ. Pathol. Toxicol. Oncol.*, 7, 159, 1986.
33. **Patel, J. M. and Block, E. R.**, Effect of NO_2 exposure on antioxidant defense of endothelial cells, *Toxicology*, 41, 343, 1986.
34. **Saladino, A. J., Willey, J. C., Lechner, J. F., Grafstrom, R. C., LaVeck, M., and Harris, C. C.**, Effects of formaldehyde, acetaldehyde, benzoyl peroxide and hydrogen peroxide on cultured human bronchial epithelial cells, *Cancer Res.*, 45, 2522, 1985.
35. **Carpenter, L. J., Johnson, K. J., Kunkel, R. G., and Roth, R. A.**, Phorbol myristate acetate produces injury to isolated rat lungs in the presence and absence of perfused neutrophils, *Toxicol. Appl. Pharmacol.*, 91, 22, 1987.
36. **Barile, F. A., Arjun, S., and Senechal, J.-J.**, Paraquat alters growth, DNA and protein synthesis in lung epithelial cell and fibroblast cultures, *ATLA*, 20, 251, 1992.
37. **Wyatt, I., Soames, A. R., Clay, M. F., and Smith, L. L.**, The accumulation and localization of putrescine, spermidine, spermine and paraquat in the rat lung: *in vitro* and *in vivo* studies, *Biochem. Pharmacol.*, 37, 1909, 1988.
38. **Ekwall, B. and Ekwall, K.**, Comments on the use of diverse cell systems in toxicity testing, *ATLA*, 15, 193, 1988.
39. **Rheinwald, J. G. and Green, H.**, Serial cultivation of strains of human epidermal keratinocytes: the formation of keratinizing colonies from single cells, *Cell*, 6, 331, 1975.
40. **Gilchrest, B. A., Marshall, W. L., Karassik, R. L., Weinstein, R., and Maciag, T.**, Characterization and partial purification of keratinocyte growth factor from the hypothalamus, *J. Cell. Physiol.*, 120, 337, 1984.
41. **Green, H.**, Cyclic AMP in relation to the proliferation of the epidermal cell: A new view, *Cell*, 15, 1801, 1978.
42. **Gilchrest, B. A., Nemore, R. E., and Maciag, T.**, Growth of human keratinocytes on fibronectin-coated plates, *Cell Biol., Int. Rep.*, 4, 1009, 1980.
43. **Johnson, E. W., Meunier, S. F., Roy, C. J., and Parenteau, N. L.**, Serial cultivation of normal human keratinocytes: a defined system for studying the regulation of growth and differentiation, *In Vitro Cell. Dev. Biol.*, 28A, 429, 1992.
44. **Wallace, K. A., Harbell, J. H., Accomando, N., Triana, A., Valone, S., and Curren, R. D.**, Evaluation of the human epidermal keratinocyte neutral red release and neutral red uptake assay using the first 10 MEIC test materials, *Toxic. In Vitro*, 6, 367, 1992.
45. **Gay, R. J., Swiderek, M., Nelson, D., and Stephens, T. J.**, The living dermal equivalent as an *in vitro* model for predicting ocular irritation, *J. Toxicol. Cutaneous Ocular Toxicol.*, 11, 47, 1992.

46. **Borenfreund, E. and Borerro, O.**, *In vitro* cytotoxicity assays: potential alternatives to the Draize ocular irritancy test, *Cell Biol. Toxicol.*, 1, 33, 1984.

47. **Borenfreund, E. and Puerner, J. A.**, Toxicity determined *in vitro* by morphological alterations and neutral red absorption, *Toxicol. Lett.*, 24, 119, 1985.

48. **Scaife, M. C.**, *In vitro* studies on ocular irritancy, in *Animals and Alternatives in Toxicity Testing*, Balls, M., Ridell, R. G., and Worden, A., Eds., Academic Press, London, 1983, 367–271.

49. **Burton, A. B. G., York, M., and Lawrence, R. S.**, The *in vitro* assessment of severe eye irritants, *Food Cosmet. Toxicol.*, 19, 471, 1981.

50. **Gettings, S. D.**, Towards validation: an update on *in vitro* eye irritation testing, *Newsletter of the Johns Hopkins Center for Alternatives to Animal Testing*, 11, 4, 1993.

Chapter 6

CELLULAR METHODS OF TERATOGENICITY

I. THE CLASSICAL ANIMAL METHODS

Standard methods for the screening of substances which are potentially toxic to the fetus are expensive and the protocols are lengthy. Chemicals which induce structural malformations, physiological dysfunction, behavioral alterations, or deficiencies in the offspring, at birth or in the immediate post-natal period, are referred to as teratogens. Most chemicals are not extensively screened for their teratogenic potential, principally because of the many toxicity tests which are already required for each chemical. Thus most pharmaceutical companies do not routinely screen therapeutic drugs for potential embryotoxicity during preclinical phases. Chemicals not destined to be used clinically are generally devoid of any teratogenic testing. This leaves the majority of substances used in the marketplace without embryotoxic data.

Classical methods for testing teratogenic substances include rodents and rabbits, principally because of their relatively short gestational periods as compared to other species (21 to 22 and 32 days, respectively). The procedures usually involve exposure of the dame either before or immediately after conception for the length of gestation. The beginning of the exposure sequence depends on the scientific questions that the investigator is interested in. For instance, if one is determining the effects of a chemical on development of the fetus, the exposure period begins after conception. If the effects on conception are desired, the exposure begins just before mating. The time sequence for the latter experiments must be carefully determined so as not to interfere with mating. Also, the trimester during which the chemical has its greatest toxic effects can be determined by exposing the gravid animal during a specific time of the gestational period. Using this protocol, answers to the mechanism of action are also ascertained.

Rodents and rabbits are also convenient because of the large size of the litters, usually about eight to ten embryos per pregnancy. This allows for the chemicals to be scored according to the number of fetuses affected. For instance, a particular indicator is selected as a measure of toxicity and the end point is then measured in each of the embryos. The minimal accepted value for this indicator is regarded as toxic and the experiment is completed in the rest of the litter. The greater the number of embryos which exhibit toxicity by this criteria, the higher the score and the more toxic the chemical. Given the potential number of chemicals that can be screened at the different dosage levels for various indicators in each fetus, the number of experiments then becomes overwhelming and expensive.

II. *IN VITRO* EMBRYOTOXICITY TESTING

A. WHOLE EMBRYO CULTURE

Another classical method is known as postimplantation embryo culture. The technique involves the removal of 1- or 2-week-old rodent embryos after which they are allowed to acclimate in culture for several hours. The viable fetuses are then exposed to different doses of chemicals and the effects on development are determined. The method is dramatic since structural changes occur quickly and teratogenicity is grossly monitored *in vitro* as growth and differentiation progress. The potential benefit of the method allows for further investigation of substances with teratogenic activity *in vivo*. In addition, culturing embryos *in vitro* provides the opportunity to study the direct action of teratogens free of maternal hormonal and pharmacokinetic influence.[1] Giavini et al.[2] have conducted postimplantation studies on the effects of ethanol and acetaldehyde on whole rat embryos of 9.5 and 10 days of gestation cultured for 48 and 30 hours, respectively. They reported that dose-related growth retardation and frequency of malformations depend on the period of gestation of the explanted embryos.

Although the procedure uses growth in culture as a component of the protocol, it should not be confused with *in vitro* cell culture methods. The embryo cultures still require extensive animal preparation and sacrifice and offer few advantages for large scale prescreening studies.

It has been suggested that *postimplantation whole embryo culture* is the best test method for screening embryotoxic substances since it represents the period of fetal development which includes morphogenesis and early organogenesis.[3] Procedures involving *preimplantation* and *early postimplantation* of embryos during organogenesis[4] (described below) may not be adequate for screening chemicals that have teratogenic potential since they monitor single cell processes which do not reflect the complex nature of embryonic development.

B. CELL CULTURE METHODS

Because of the problems associated with traditional embryotoxicity testing, efforts have been made to develop *in vitro* teratogenicity tests. A prerequisite for the use of embryonic tissue in culture is that the isolated organs or cells should be maintained in culture for at least 48 to 72 hours. Several methods, incorporating either isolated cells or organs, have been introduced recently. A "micromass" method has been described[5] for the culture of dissociated arm and leg rudiment cells. These cells are capable of synthesizing a variety of extracellular matrix proteins which are incorporated into basement membranes. The production of these proteins is inhibited by most known teratogenic chemicals and they can be measured by classical spectrophotometric analysis or by radioactive labeling with amino acid precursors.

1. Preimplantation Methods

The first description of the culture of *preimplantation* embryos was reported by Brachet[6], and the method has since shown its importance in experimental embryology and reproductive biology. The technique is described for the preimplantation culture of embryos from mouse, rabbit, and human. In general, 3- to 6-day-old embryos (3-day morulae and 4-day blastocysts) are removed from superovulated sexually mature animals and transplanted into culture dishes containing conventional media or uterine flushing-supplemented media to which is added bovine serum albumin, serum, or serum-free conditioned medium.[7-9] The cultures are exposed to the test substance during the incubation period and are examined ultrastructurally or monitored for growth and proliferation using biochemical procedures. Laschinski et al.[10] evaluated the cytotoxicity of xenobiotics in cultured embryonal stem cells derived from mouse blastocysts. Using the kenacid blue, neutral red, and MTT assays, they showed that the embryonal cells were more sensitive to known teratogens than fibroblast cultures.

Thus, preimplantation embryo cultures may provide an interesting system for testing developmental toxicity after exposure to xenobiotics.

2. Continuous Cell Culture Methods

The use of finite or permanent continuous cells lines are used for monitoring teratogenicity *in vitro*. Braun and Horowicz[11] tested 32 pesticides for teratogenicity in a permanent embryonic cell line by measuring cell attachment. Pratt and Willis[12] measured growth inhibition in human embryonic cells as a screening technique for teratogens. Barile et al.[13,14] and Barnes et al.[15] describe experiments using radiolabeled proteins, synthesized by human fetal lung fibroblasts, as an indicator of cellular toxicity (Figure 1). The cells are derived from a first trimester fetus and represent the period of gestation when the embryo is most susceptible to developmental anomalies. The fibroblasts are easily maintained in continuous culture and exhibit a population doubling level of 50 ± 10. Other human fetal lung cells, such as MRC-5 and WI38, have also been extensively studied and characterized in aging studies.[16,17] It must be emphasized, however, that these continuous cell lines may not be suitable cell culture models for screening chemicals with potential toxicity to fetal *development*, since they are beyond the primary culture stage. Consequently, their response may actually reflect basal cytotoxicity rather than a specific effect on the growing fetus.

3. Organ Culture Methods

Fetal organs have been maintained in culture in the primary cell stage by explanting the organ and partially or totally submerging the tissue in growth medium. The tissue explant is placed on a micron filter or on plastic coated with extracellular matrix components. This approach has been useful in the culture of limb buds from several species of animal in serum containing- or chemically defined-medium.[18] Morphological development and differentiation is monitored, as well as biochemical parameters. Renal development and suspected nephrotoxic agents have been investigated in organ cultures of mouse kidney.[19] The isolation

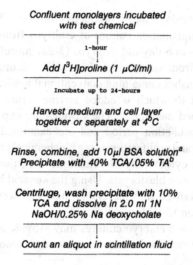

Confluent monolayers incubated
with test chemical

1-hour

Add [³H]proline (1 μCi/ml)

Incubate up to 24-hours

Harvest medium and cell layer
together or separately at 4°C

Rinse, combine, add 10μl BSA solution[a]
Precipitate with 40% TCA/.05% TA[b]

Centrifuge, wash precipitate with 10%
TCA and dissolve in 2.0 ml 1N
NaOH/0.25% Na deoxycholate

Count an aliquot in scintillation fluid

Figure 1. Outline of [³H]proline incorporation procedure. [a] Indicates bovine serum albumin, 1 mg/ml. [b] Indicates trichloroacetic acid, tannic acid solution, store at 4°C.

of primary organotypic cultures of type II epithelial cells from fetal rat lung has also been realized.[20,21] These specialized cells are capable of synthesizing and storing surfactant, which is composed of phospholipids and is secreted into the alveolar space. *In vivo*, surfactant coats the alveolar walls and decreases the surface tension present at the air-cell interface and thus allows for oxygen to penetrate the thin alveolar structures. Consequently, the cells may be used in toxicity testing studies to monitor embryotoxic effects in the lungs, particularly in relation to surfactant production and secretion. Thus, as with other organ specific primary cultures, their advantage lies in their ability to act as a model for detecting target organ toxicity.

In conclusion, the protocols and procedures described are currently undergoing extensive testing by industrial concerns and are being subjected to interlaboratory validation. Thus a variety of methods may be included in a battery of tests for testing chemicals with embryotoxic potential.

REFERENCES

1. **Steele, C. E., New, D. A. T., Ashford, A., and Chopping, G. P.,** Teratogenic action of hypolipidemic agents: an *in vitro* study with postimplantation rat embryos, *Teratology*, 28, 229, 1983.
2. **Giavini, E., Broccia, M. L., Prati, M., Bellomo, D., and Menegola, E.,** Effects of ethanol and acetaldehyde on rat embryos developing *in vitro*, *In Vitro Cell. Dev. Biol.*, 28A, 154, 1992.
3. **Giavini, E., Broccia, M. L., and Prati, M.,** Teratogenicity testing in vitro: post-implantation whole-embryo culture, *ATLA*, 119, 94, 1991.
4. **Sadler, T. N.,** Culture of early somite mouse embryos during organogenesis, *J. Embryol. Exp. Morphol.*, 49, 17, 1979.

5. **Flint, O. P.**, An *in vitro* test for teratogens using cultures of rat embryo cells, in *In Vitro Methods in Toxicology*, Atterwill, C. K. and Steele, C. E., Eds., Cambridge University Press, London, 1987, 339–364.

6. **Brachet, A.**, Recherches sur le determinism hereditaire de l'oeuf des mammiferes. Development *in vitro* de jeunes vesicules blastodermique du lapin, *Arch. Biol.*, 28, 447, 1913.

7. **Fischer, B., Lambertz, M., and Hegele-Hartung, C.**, Ultrastructural and autoradiographic study of preimplantation rabbit embryos grown in conventional or uterine flushing-supplemented culture media, *In Vitro Cell. Dev. Biol.*, 28A, 337, 1992.

8. **McKiernan, S. H. and Bavister, B. D.**, Different lots of bovine serum albumin inhibit or stimulate in vitro development of hamster embryos, *In Vitro Cell. Dev. Biol.*, 28A, 154, 1992.

9. **Kobayashi, K., Takagi, Y., Satoh, T., Hoshi, H., and Oikawa, T.**, Development of early bovine embryos to the blastocyst stage in serum-free conditioned medium from bovine granulosa cells, *In Vitro Cell. Dev. Biol.*, 28A, 255, 1992.

10. **Laschinski, G., Vogel, R., and Spielmann, H.**, Cytotoxicity test using blastocyst-derived euploid embryonal stem cells: a new approach to *in vitro* teratogenesis screening, *Reproduc. Toxicol.*, 5, 57, 1991.

11. **Braun, A. G. and Horowicz, P. B.**, Lectin-mediated attachment for teratogens: results with 32 pesticides, *J. Toxicol. Environ. Health*, 11, 275, 1983.

12. **Pratt, R. W. and Willis, W. D.**, *In vitro* screening assay for teratogens using growth inhibition of human embryonic cells, *Proc. Natl. Acad. Sci. U.S.A.*, 82, 5791, 1985.

13. **Barile, F. A., Ripley-Rouzier, C., Siddiqi, Z. and Bienkowski R. S.**, Effects of puromycin and hydroxynorvaline on net production and intracellular degradation of collagen in human fetal lung fibroblasts, *Arch. Biochem. Biophys.*, 270, 294, 1989.

14. **Barile, F. A., Guzowski, D. E., Ripley, C., Siddiqi, Z., and Bienkowski, R.S.**, Ammonium chloride inhibits basal degradation of newly synthesized collagen in human fetal lung fibroblasts, *Arch. Biochem. Biophys.*, 276, 125, 1990.

15. **Barnes, Y., Houser, S., and Barile, F. A.**, Temporal effects of ethanol on growth, thymidine uptake, protein, and collagen production in human fetal lung fibroblasts, *Toxic. In Vitro*, 4, 1, 1990.

16. **Hayflick, L.**, The limited *in vitro* lifetime of human diploid cell strains, *Exp. Cell Res.*, 37, 614, 1965.

17. **Jacobs, J. P., Jones, C. M., and Baillie, J. P.**, Characteristics of a human diploid cell designated MRC-5, *Nature (London)*, 227, 168, 1976.

18. **Beck, F. and Gulamhusein, A. P.**, The contrast between mouse and ferret limb buds in culture — possible advantages of comparing results from a limb culture system with a whole embryo explantation, in *Teratology of the Limbs*, Merker, H.-J., Nau, H., and Neubert, D., Eds., Walter de Gruyter, Berlin, 1980, 117–127.

19. **Saxen, L. and Ekblom, P.**, The developing kidney as a model system for normal and impaired organogenesis, in *Culture Techniques. Applicability for Studies on Prenatal Differentiation and Toxicity*, Neubert, D. and Merker, H.-J., Eds., Walter de Gruyter, Berlin, 1981, 291–300.

20. **Shami, S. G., Aghajanian, J. D., and Sanders, R. L.**, Organotypic culture of fetal lung type II alveolar epithelial cells: applications to pulmonary toxicology, *Environ. Health Perspect.*, 56, 87, 1984.

21. **Post, M., Torday, J. S., and Smith, B. T.**, Alveolar type II cells isolated from fetal rat lung organotypic cultures synthesize and secrete surfactant-associated phospholipids and respond to fibroblast-pneumonocyte factor, *Exp. Lung Res.*, 7, 53, 1984.

Chapter 7

CELLULAR METHODS OF IMMUNE FUNCTION

I. INTRODUCTION TO IMMUNOCYTOTOXICITY TESTING

The increasing realization of the role that the immune system plays in the overall screening of environmental pollutants and the impact of these chemicals on its function, has led to important advances in the understanding of cellular immunity. In general, these advances were prompted by the same impulses observed in studies of other organ systems — that is, by the inclination to understand the mechanisms of toxicity as they involved cells of immune function. Thus mechanistic studies have promoted the biomedical advances that toxicologists are now exploring. These advances have been initiated in part by studies in immunology and molecular biology which have made available the precise techniques necessary to investigate the role of these cells *in vivo*. Our understanding of the field burgeoned with the isolation of primary cultures of specialized immune cells. The application for toxicity testing, therefore, lies in the ability of xenobiotics to alter the known responses generally associated with these cells. The development of *in vitro* methods for evaluating toxicity to the immune system has progressed by attempts to establish these cells as models for toxicity testing. Thus, the evolution of many of the procedures involved with screening of potential immunotoxins is analogous to similar developments in other organ systems. The sounding board for the arguments for or against these model systems may appear to be the same as those described above. This section, therefore, will limit the discussion to the application of the function of the immune response to its capacity for handling xenobiotic substances and its adaptation to *in vitro* cytotoxicity testing.

II. STRUCTURE AND FUNCTION OF THE IMMUNE SYSTEM

Immunology is the science concerned with the processes by which humans and animals defend themselves against infection (thus the science of *immunity*). (It should be noted, however, that the term "immunity" implies resistance to infectious agents, foreign particles, bacteria, fungi, viruses, and transformed mammalian cells, as well as toxins and chemicals.) Two types of immunity are recognized: (1) natural or innate immunity; and (2) acquired or specific immunity.

A. NATURAL IMMUNITY
Natural immunity involves the resistance of the host to foreign substances without necessarily triggering a cascading series of events terminating in

107

destruction of the invading material. This "natural" defense includes the concept of hereditary susceptibility, the presence of epithelial barriers (such as the skin and lungs), the vascular inflammatory response (hyperemia and the formation of fluid exudate), complement, phagocytosis (chemotaxis, opsonization, degranulation), the acidic pH of the stomach, and the normal intestinal flora. It is important to recognize that although these systems function primarily as a result of their programmed genetic predisposition, they have some influence on the operations of acquired immunity. For instance, the processes which trigger phagocytosis stimulate macrophages to act accordingly. Also, the vascular response to inflammation usually leads to cellular reactions involving leukocytes. Thus the systems are not necessarily distinct entities, but may act collectively to deter invading substances.

B. ACQUIRED IMMUNITY

Acquired immunity involves the adaptive response to foreign antigenic stimuli, resulting in the acquisition of immunologic memory and the production of antibodies. Acquired immunity may be further classified according to these functions, which are interrelated: (1) *humoral immunity* and (2) *cell-mediated immunity*. Humoral immunity refers to the synthesis and release of free antibody into the vascular and lymphatic circulations. The antibody refers to a group of plasma proteins (immunoglobulins) which react specifically with the antigen eliciting the reaction (also known as the phenomenon of *specificity*). Cell-mediated immunity involves the stimulation of "sensitized" lymphocytes, which subsequently initiate the destruction of chemical or cellular invading substances.

Several organs constitute that portion of the immune system required for activating acquired immunity, all of which are important in the cohesive response of the body to chemical invasion. The bone marrow, thymus gland, lymph nodes, and spleen participate in the uptake and recognition of foreign matter, in the subsequent transfer of the information necessary for recognition of the antigen by other immune cells, or in the destruction of the substance itself.

The cellular elements act independently and cohesively, to maintain an integrated pattern of surveillance to closely monitor cellular and chemical intrusions into the body. Thus the mammalian system involves multiple interactions between numerous subsets of thymus-derived lymphocytes (T cells), bone marrow-derived lymphocytes (B cells), and other accessory cells originating from macrophages, including polymorphonuclear leukocytes, natural killer cells (NK cells), neutrophils, basophils, eosinophils, and mast cells. Table 1 summarizes the major cell types involved in cell-mediated immunity and their normal functions. The advances in cytotoxicity testing have resulted primarily from the ability to culture only a few of these cell types, namely, primary and continuous cultures of T cells, natural killer cells, and some cells of macrophage origin.

III.   NATURAL KILLER CELLS IN *IN VITRO* TOXICOLOGY

The ability of some cells of the immune system to be spontaneously activated upon a cytotoxic insult is an important mechanism in natural resistance to tumors and viral diseases. Natural killer (NK) cells are active in tumor surveillance[1,2,12] and can limit the progress of viral infections.[3-5] In addition, some viruses, malarial parasites, bacteria, and airborne fungi have shown sensitivity to NK cell activity or have provoked an elevation in NK cell activity *in vivo*.

NK cells have neither T cell receptor for antigen (designated TCR) nor CD3 (an important complement protein necessary for initiation of cytotoxic action). Thus they are not true T cells and their activity does not appear to represent a typical immune phenomenon. The only specificity associated with them is their broad range of cultured target cells, most important of which are tumor-derived cells. In addition, no evidence of specific memory response exists, although activity against tumor cells can be enhanced by prior sensitization to any particular type of tumor.[4-6]

NK cells have receptors for the F_c portion of immunoglobulin G (IgG) and therefore can mediate antibody-dependent cell-mediated cytotoxicity (ADCC).[7] This receptor has been used as a marker for NK cells and as a means of identification, although some large granular lymphocytes (also known as K or killer cells, Table 1) may also be positive for the marker in monoclonal antibody assays.[8] NK cells have the ability to act as large granular lymphocytes. Also, like cytotoxic T lymphocytes (CTL cells) and K cells, NK cells have cytoplasmic azurophilic granules containing perforin.[9-11] This protein is encountered in cytoplasmic granules of these cells as well as in eosinophils, and are capable of forming channels in the target cell membrane, thus rendering it susceptible to influx of ions and small molecules and subsequent cell lysis. Consequently, NK cells possess cytotoxic mechanisms which are similar to other T cells.

A. ISOLATION AND COLLECTION OF NK CELLS

Peripheral lymphoid cells consist of monocytes and a heterogeneous assortment of lymphocytes. The collection of NK effector cells is usually performed during the separation of human peripheral blood monocytes using separation medium.[9,10] Blood samples are collected, diluted, and layered on lymphocyte separation medium (LSM). The suspension is centrifuged and the monocyte band is removed, diluted, and recentrifuged. The cells are then washed and plated onto plastic dishes precoated with fetal calf serum. Monocytes preferentially adhere to the plastic after 1 h, leaving the nonadherent NK cells available for study in the overlying suspension.

B. MONITORING NK CELL ACTIVITY

NK tumoricidal activity is assessed in a classical 4-h microcytotoxicity assay by culturing splenocytes or peripheral blood lymphocytes with a ^{51}Cr- or

TABLE 1
Major Cell Types Associated with Cell-Mediated Immunity and Their Functions *In Vivo*

Cell type[a]	Classification	Functions
NK cell	Killer cell	Tumor surveillance; inhibit viral, bacterial, and fungal progress
LAK cell	Killer cell	Tumor surveillance and destruction of tumor cells
Macrophages	Mononuclear phagocyte	Phagocytosis, lymphocyte activation, and secretion of immunomodulators
Neutrophils	Polymorphonuclear phagocyte	Phagocytosis; participate in the inflammatory response and chemotaxis
CTL	Thymus-derived T Lymphocytes	Cell-mediated immunity; particularly to viral and neoplastic invasion, as well as antigenic response
Basophils, mast cells	Circulating leukocytes	Inflammatory response; release vasoactive amines

[a] NK = natural killer cell; LAK = lymphokine-activated killer cell; CTL = cytotoxic T lymphocyte.

fluorochrome-labeled cell line sensitive to the lytic action of NK cells.[7,11-16,50] The cell lines used are typically murine YAC-1 or human K562 lymphoma cells. Known proportions of NK effector cells and target cells are mixed and incubated. An important control of the assay must incorporate multiple effector/target cell ratios in order to accurately gauge the concentration-effect response between the two cell lines. The degree of cell lysis of the target cells is measured by the corresponding release and measurement of radioisotope or fluorescence in the supernatant as target cells are destroyed, thus reflecting NK cell activity. Experimental protocols have been extensively detailed in several reviews.[15-19]

Several techniques are currently available for measuring NK cell activity in response to the action of cytotoxic chemicals. These procedures were developed primarily to ascertain the mechanism of NK activity, but are used in immunocytotoxicity testing. Such techniques as the single cell assay in agarose,[20,21] limiting dilution analysis,[22] and the modified plaque-forming assays[23] have allowed investigators to determine the frequency and the recycling potential of actively lytic cells, that is, the number of target cells destroyed by effector cells. The assays have been applied to measuring the kinetics of cell-mediated toxicity and the effects of chemicals on NK cell activity.

IV. MONOCYTES/MACROPHAGES IN
IN VITRO TOXICOLOGY

A. MACROPHAGES

Macrophages are one of the major classes of mononuclear phagocytes (monocytes) originating from the bone marrow. Fixed macrophages, such as Kupfer cells in the liver, do not migrate, whereas alveolar macrophages are motile and circulate principally in the capillaries of the lungs and spleen. Both types of cells are 10 to 20 μm in diameter and are distinguished from the other major mononuclear phagocytic cell type, neutrophils, by their unlobed nucleus, absence of specific neutrophilic granules, and the presence of microvilli.

Both fixed and motile macrophages play an important role in the immune response and resistance to infections and neoplastic invasion.[24,25] They are actively phagocytic, engulfing foreign particles and cellular debris into intracytoplasmic vacuoles. Phagocytosis is preceded by the binding of the foreign particle to the cell surface, especially material which has been opsonized (coated with specific antibody). The antibodies which have adhered to the membrane antigen attach to the F_c receptor located on the macrophage cell membrane, as well as receptors for complement component C3b. The F_c and C3b receptors on the membranes of the macrophages are revealed in culture during the progress of cytotoxicity assays by their rosette formation with antibody-coated red blood cells. Once the foreign particle is ingested, the vacuoles traverse the cytoplasm and fuse with lysosomes, and the contents are digested by a variety of hydrolases.

Macrophages also play an important role in the activation of lymphocytes in humoral and cell-mediated immunity. Fragments of ingested antigen are presented at the cell surface and stimulate T-helper cells as part of cell-mediated response. Cytokines (some of which are known as lymphokines) are a group of soluble peptide cell regulators secreted by a variety of cells, as well as lymphocytes, and activate macrophage phagocytic activity. These cytokines include the interferons (IFN-α, β, and τ),[26] interleukin-2 (IL-2),[27] interleukin-3 (IL-3),[28,29] and interleukin-4 (IL-4).[26,30] Other lymphokines, such as macrophage chemotactic factor (MCF) and macrophage activating factor (MAF) also mediate macrophage activity. MCF is released by sensitized lymphocytes and stimulates migration of macrophages and monocytes. Similarly, MAF is secreted by sensitized lymphocytes and activates macrophages to increase their phagocytic activity, protein synthesis, and cell surface motility. Macrophages have also been shown to produce prostaglandins and thromboxane B_2, which are speculated to play a role in immunosuppression.[31]

B. ISOLATION AND COLLECTION OF MACROPHAGES

Macrophages from the peritoneal cavity are isolated and collected primarily from mice.[32,33] The cells are induced by injecting sterile thioglycollate broth intraperitoneally. Five days later, cells are collected by peritoneal lavage pyrogen-free saline, washed, and suspended in growth medium (RPMI is commonly used) containing 10 to 20% fetal calf serum. Cells are then incubated for 3 h at 37°C in a humidified chamber containing 5 to 10% CO_2, after which the cultures are washed to remove nonadherent cells.

Alveolar macrophages are obtained from rodents by pulmonary lavage.[34] Rats or mice are anesthetized and 10 10-ml aliquots of Ca^{2+}- and Mg^{2+}-free Hanks' balanced salt solution (HBSS) are instilled separately in the trachea. After each injection, the fluid is aspirated and collected, and the combined aspirates are concentrated with low speed centrifugation. The pelleted cells are resuspended in buffered medium and incubated as above.

Cultures are identified morphologically, by light or electron microscopy, and cytochemically using qualitative and quantitative methods for the demonstration of nonspecific esterases.[35]

C. MONITORING MACROPHAGE ACTIVITY FOR TOXICITY TESTING

In vitro studies on macrophages have been applied to immunocytotoxicity testing, primarily due to their ease of isolation and their ability to be cultured. Most of these tests have included the measurements of both a cytotoxic response directly on the cultured cells and the ability of the differentiated macrophages to maintain active phagocytic properties in the presence of chemical agents. In addition, the effects and influence of cell mediators, such as cytokines and prostaglandins, have been used as analytical probes or as measurable indicators in cytotoxic techniques.

Numerous assays are available for measuring the cytotoxic influence of chemicals on various functions performed by macrophages.[36–41] These procedures were developed principally to study the immunologic properties of macrophages and monocytes *in vitro* and *in vivo*, and have since been effectively incorporated into the toxicological arena.

The development of flow cytometry has rapidly advanced toxicological investigations of substances capable of interfering with macrophage activity. The technique is based on the ability of subpopulations of macrophages and mononuclear cells to actively phagocytize particles, which are later separated according to their phagocytic properties and cell size. Essentially, cell suspensions or tissue aggregates are incubated with fluorescent microspheres or monoclonal antibodies, specific for cell surface antigens, conjugated to different fluorescent dyes. The suspension is passed through the cell sorter which separates the emerging stream of cells into about 40,000 droplets per second. A laser is directed at the stream, and fluorescence is detected by the electronic photocells. Green fluorescence is perceived as a function of particle voltage, red fluorescence as a measure of DNA, and forward scattered light is registered according to cell size. If the signals from the photocells meet the criteria for fluorescence and size, as set by the operator, the cells retain an electrical charge and are sorted into different collecting tubes based on these electronically assigned charges. The procedure can accurately and sensitively quantitate various cellular events through the simultaneous measurement of multiple parameters on many cells in seconds.[42-45] Flow cytometry provides useful data regarding the toxic effects upon cell cycle phases, including concurrent determination of DNA and RNA synthesis,[46] cellular RNA content,[47] and the determination of cell surface activation markers.[48] In addition, the protocol allows for the physical and statistical separation of specific cell types based on their surface and cytoplasmic phagocytic markers.[42] Details of the procedures for specific applications have been reviewed.[45]

ADCC is another frequently used method for assessing macrophage activation and binding.[24,49] The procedure relies on the ability of the effector cells (typically macrophages or mononuclear cells) to recognize and destroy target cells (erythrocytes or tumor cell lines) by binding to specific antibodies on the surface of the targets. The target cells are labeled with a radioactive marker (^{51}Cr, ^{125}I), and the degree of macrophage activity is monitored by the measurable release of radioactivity following target cell death. ADCC has been used to investigate the effect of various substances on macrophage activation and cytotoxicity, including lipopolysaccharide,[32] endotoxin,[33] and morphine.[51]

Macrophage activity has also been determined according to their response in the presence of soluble mediators, such as MAF and MCF, and their ability to secrete prostaglandins.[52]

V. NEUTROPHILS IN *IN VITRO* TOXICOLOGY

Neutrophils are chemotactic phagocytic polymorphonuclear cells (PMNs) that ingest bacteria, foreign particulate matter, and damaged tissue cells *in vivo*. They are the most abundant of the circulating white blood cells, are the dominant cell type in the inflammatory reaction, and are activated in response to C5a complement protein. They are mobilized quickly to sites containing infectious agents and antigen-antibody complexes. *In vitro* they respond to several kinds of attractants, including bacterial culture fluids and zymosan activated serum. Their usefulness in cytotoxicity testing studies has focused on their chemotactic ability,[53,54] lysosomal enzyme release,[55] and their phagocytic response in the presence of *Staphylococcus aureus*.[56]

As with eosinophils, neutrophils have similar buoyant densities to macrophages (1.085 to 1.095 g/ml) under isotonic conditions. These densities tend to overlap to some extent with the density of erythrocytes at about 1.09 to 1.11 g/ml. A single-step method for the isolation of PMNs, predominantly consisting of neutrophils, uses a solution of sodium metrizoate and the polysaccharide, dextran 500.[57] The separation relies on the high osmolarity of the solution

(450 mOsm) resulting in dehydration of the erythrocytes and the formation of a short density gradient beneath the sample-medium interface. The increased density of the erythrocytes allows them to sediment through the gradient (1.11 g/ml), while the PMNs migrate to a position about 1 cm below the interface. The dextran also increases the sedimentation rate of the red cells. The result is pelleting of the red blood cells and the formation of a band of PMNs at a position below the less dense mononuclear cells. It is important that the time, speed, and temperature of the separation procedure are carefully monitored since PMNs continuously lose water during the centrifugation steps.

VI. LYMPHOCYTES IN *IN VITRO* TOXICOLOGY

A. LYMPHOCYTES AND CELL-MEDIATED IMMUNITY

During the past decade, many studies have examined the functional aspects of T lymphocytes involved in cell-mediated immunity. The *in vitro* techniques include lymphocyte proliferation, T cell cytotoxicity, and lymphokine production. Delayed hypersensitivity techniques are also commonly employed for *in vivo* clinical and experimental assessment.

The measurement of lymphoproliferative responses is frequently used to study cell-mediated cytotoxicity. In the blastogenic (or transformation) assays, lymphocytes are induced *in vitro* with mitogens such as phytohemagglutinin (PHA), concanavalin A (ConA), and pokeweed mitogen (PWM) to selectively stimulate T lymphocytes. Bacterial products, such as lipopolysaccharide (LPS), stimulate B lymphocytes. These mitogens activate a high percentage of T cells to undergo DNA, RNA, and protein synthesis in a polyclonal fashion, as well as to stimulate membrane phospholipid synthesis, nutrient transport, and glycolysis.[64,65] To evaluate this response, suspensions of isolated lymphocytes are incubated with mitogens for 72 h, after which [^{3}H]thymidine is added to the tubes or cultures for an additional 24 h. The trichloroacetic acid precipitable radioactivity is measured as an indicator of DNA synthesis.

The induction and proliferation of CTL represents an important acquired effector mechanism of cell-mediated immunity, particularly in response to viral and neoplastic invasion *in vivo*. Cytotoxic T cells can be induced *in vivo* by immunization of animals with allogeneic lymphocytes or target tumor cells (described below) or *in vitro* in mixed lymphocyte cultures. In contrast to NK cells, the specificity of the attack of CTL is mediated by the α/β T cell receptor (TCR) for antigen on the lymphocyte. CTL respond only to cell surface antigens, particularly those produced by virus-infected cells, malignant cells, allogeneic cells, and cells expressing fragments of conventional protein antigens or haptens. The measurement of cytolytic activity or response to chemicals can be assayed with techniques including radioactive labeling of target cells with ^{51}Cr. Direct cytotoxicity to the cytotoxic T cells is measured by their ability to contact and destroy target cells after labeling. The degree of lysis is monitored by release of ^{51}Cr from the target cell within 3 to 24 h.[59]

Mature human T lymphocytes also carry CD2 receptors that bind to determinants on the sheep red blood cell (SRBC) surface. When mixed with SRBCs, T cells bind them to their surface forming "E" (erythrocyte) rosettes. The formation of rosettes is a convenient method for identifying and counting human T cells, for separating T cells from B cells, and for monitoring T cell activity. Any inhibition of rosette formation by xenobiotics suggests interference with glycolytic, protein, or nucleic acid metabolic pathways.[60,61]

Lymphokines are soluble mediators synthesized and secreted by active lymphocytes and act as intercellular messengers of cell-mediated immunity[62] (see Section IV, Macrophages, in this chapter). Some of the lymphokines are now detectable by radioimmunoassay and are used to assess cell-mediated cytotoxicity.[63]

In vitro procedures involving T cells and the parameters used to evaluate their functions are used to study the mechanisms of toxicity to a variety of toxins and drugs, including PCBs,[66] marijuana,[67-69] morphine,[70] cocaine,[71] and polycyclic aromatic hydrocarbons.[72,73]

B. ISOLATION AND COLLECTION OF LYMPHOCYTES

The standard method of isolating lymphocytes from human peripheral blood, along with the isolation of other monocytes, was developed by Boyum.[58] Human peripheral blood is collected in anticoagulant and diluted with an equal volume of isotonic saline. The blood suspension is layered over a solution consisting of a mixture of sodium metrizoate and Ficoll® 400, with a density of 1.077 g/ml and an osmolality of 295 mOsm. After a short low speed centrifugation, the mononuclear cell fraction is harvested from the sample-medium interface. Recently, other density gradients have been developed (Nycodenz®) that allow them to be combined with undiluted blood samples. The mononuclear cells float toward the top of the mixture and other cells are pelleted. Recovery and viability of the harvested mononuclear cells are at least equivalent to those of standard methods, and the method eliminates the tedious layering of blood.[74] Mononuclear cells are then removed, suspended in saline, and centrifuged to remove platelets. Lymphocytes are separated from monocytes by resuspending in RPMI 1640 medium containing 10% fetal calf serum. Monocytes settle and adhere after 1 h in culture at 37°C, while the lymphocytes are removed with the overlying medium since they remain in suspension.

CTL are also isolated from immunized mice. Approximately 7 to 10 days prior to isolation, mice are injected with viable target tumor cells for the production of immune lymphocytes. The isolation of splenic lymphocytes from pooled spleens and peritoneal exudate lymphocytes from peritoneal washings of mice has been described.[75]

C. CONTINUOUS LYMPHOCYTE CULTURES

The American Type Culture Collection (ATCC, Bethesda, MD) has many cell lines of human and animal origin as part of the Tumor Immunology Cell

Bank. Many of the cell lines, including continuous cultures of myelomas, hybridomas, and lymphomas, have been characterized according to their immunologic properties. Cell surface markers are identified, and sensitivity or resistance to PHA, cortisol, thioguanine, and thymidine is noted, as well as any response to or stimulation by cytokines.

It is possible that the mechanisms which underlie immunotoxic phenomena may be explained by interaction of chemicals with the basal cytotoxic components of the cell. In essence, the same mechanisms which have been used to detect general systemic toxicity in established cell lines may also play a critical role in describing the reaction of lymphocytes and macrophages to these chemicals. Although differentiated B and T cells are pharmacologically distinguishable, they may in fact react the same in the presence of most cytotoxic compounds. Consequently, a cell test battery consisting of acute general toxicity testing procedures could be formulated to test for immunotoxicity. It may be necessary, however, to await the results of further *in vitro* immunocytotoxicity testing studies, involving lymphocytes and macrophages, before our understanding of the contribution of immunological procedures to general toxicity testing protocols expands.

VII. TIER TESTING IN MICE

The immunotoxicology laboratories of the National Institute of Environmental Health Sciences (NIEHS, National Institutes of Health) and National Toxicology Program (NTP) have reported on the design of a screening battery involving a "tier" approach for detecting potentially immunotoxic compounds in mice.[76,77] Tier I was designed to include tests to screen for toxicants which do not induce overt toxicity, such as assays for cell-mediated immunity, humoral immunity, and immunopathology. Tier II includes tests which further the understanding of the immunotoxic phenomena, as well as assays to monitor splenic lymphocyte subpopulations and host resistance. A database was generated in an attempt to improve future testing strategies in the quantitative risk assessment for immunotoxicity. The results suggest that only two or three immune tests are sufficient to predict immunotoxic compounds in rodents. In addition, the relationship between immunotoxicity, carcinogenicity, and genotoxicity is explained.[78-80]

REFERENCES

1. **Introna, M. and Mantovani, A.,** Natural killer cells in human solid tumors, *Cancer Metastasis Rev.*, 2, 337, 1987.
2. **Hanna, N.,** Role of natural killer cells in control of cancer metastasis, *Cancer Metastasis Rev.*, 1, 45, 1982
3. **Biron, C. A. and Welsh, R. M.,** Activation and role of natural killer cells in virus infections, *Med. Microbiol. Immunol.*, 170, 155, 1983.

4. **Herberman, R. B.**, Introduction, in *Mechanisms of Cytotoxicity of NK cells*, Herberman, R. B. and Callewaert, D. M., Eds., Academic Press, Orlando, 1985, chapter 1.
5. **Riccardi, C., Santoni, A., Barlozzari, T., Puccetti, P., and Herberman, R. B.**, In vivo natural reactivity of mouse against tumor cells, *Int. J. Cancer*, 25, 475, 1980.
6. **Abrruzo, L. V. and Rowley, D. A.**, Homeostasis of the antibody response: immunoregulation by NK cells, *Science*, 222, 581, 1983.
7. **Wright, S. C. and Bonavida, B.**, Studies on the mechanism of natural killer cell-mediated cytotoxicity (CMC). I. Release of cytotoxic factors specific for NK cells-sensitive target cells (NK cells CF) during co-culture of NK cells effector cell with NK cells target cells, *J. Immunol.*, 129, 433, 1982.
8. **Riccardi, C., Puccetti, P., Santoni, A., and Herberman, R. B.**, Rapid *in vivo* assay of mouse natural killer cell activity, *J. Natl. Cancer Inst.*, 63, 1041, 1979.
9. **Timonen, T. and Saksela, E.**, Isolation of human killer cells by density gradient centrifugation, *J. Immunol. Methods*, 36, 285, 1980.
10. **Timonen, T., Ortaldo, J. R., and Herberman, R. B.**, Characteristics of human large granular lymphocytes and relationship to natural killer and K cells, *J. Exp. Med.*, 153, 569, 1981.
11. **Wright, S. C. and Bonavida, B.**, Selective lysis of NK cells-sensitive target cells by a soluble mediator released from murine spleen cells and human peripheral blood lymphocytes, *J. Immunol.*, 126, 1516, 1981.
12. **Kiessling, R., Klein, E., Pross, H., and Wigzell, H.**, Natural killer cells in the mouse. I. Cytotoxic cells with specificity for mouse Moloney leukemia cells. Specificity and distribution according to genotype, *Eur. J. Immunol.*, 5, 112, 1975.
13. **Darzynkiewiez, Z. and Andreff, M.**, Multiparameter flow cytometry. I. Application in analysis of the cell cycle, *Clin. Bull.*, 11, 47, 1981.
14. **Pollack, A., Bagwell, B., Irwin, G., and Jensen, J.**, The kinetics of the formulation of a G2 block from tritiated thymidine in phytohemagglutin stimulated human lymphocytes, *Cytometry*, 1, 57, 1980.
15. **Pillai, R. M. and Watson, R. R.**, In vitro immunotoxicology and immunopharmacology: studies on drugs of abuse, *Toxicol. Lett.*, 53, 269, 1990.
16. **Dean, J. H., Luster, M. I., Boorman, G. A., and Lauer, L. D.**, Procedures available to examine the immunotoxicity of chemicals and drugs, *Pharmacol. Rev.*, 34, 137, 1982.
17. **Krakowka, S.**, In vitro and in vivo methods in immunotoxicology, *Toxicol. Pathol.*, 15, 276, 1987.
18. **Dean, J. H., Luster, M. M. I., and Boorman, G. A.**, Methods and approaches for assessing immunotoxicity: an overview, *Environ. Health Perspec.*, 43, 27, 1982.
19. **Dean, J. H. and Thurmond, L. M.**, Immunotoxicology: an overview, *Toxicol. Pathol.*, 15, 265, 1987.
20. **Grimm, E. and Bonavida, B.**, Mechanism of cell mediated cytotoxicity at the single cell level. I. Estimation of cytotoxic T lymphocyte frequency and relative lytic efficiency, *J. Immunol.*, 123, 2861, 1979.
21. **Grimm, E., Thomas, J. A., and Bonavida, B.**, Mechanism of cell-mediated cytotoxicity at the single cell level. II. Evidence for first order kinetics of T cell mediated lysis and for heterogeneity of lytic rate, *J. Immunol.*, 123, 2870, 1979.
22. **Pross, H. F., Wilberforce, K., Dean, D., Kennady, J. C., and Baines, M. G.**, Spontaneous lymphocyte mediated cytotoxicity against tumor cells. VIII. Characteristics of normal lymphocytes showing consistently high or low natural killer (NK) cell activity, in *The Molecular Basis of Immune Cell Function*, Kaplan, J. G., Ed., Elsevier Press, Amsterdam, 1989, 580–600.
23. **Bonavida, B., Ikerjiri, B., and Kedar, E.**, Direct estimation of frequency of cytotoxic T lymphocytes by a modifed plaque assay, *Nature*, 263, 769, 1976.
24. **Snyderman, R., Pike, M. C., Fischer, D. G., and Koren, H. S.**, Biologic and biochemical activities of continuous macrophage cell lines P388D$_1$ and J774, *J. Immunol.*, 119, 2060, 1977.

25. **Sorrell, T. C., Lehrer, R. I., Ferrari, L. G., Muller, M., and Selsted, M. E.**, Divergent expression of cytotoxic and microbicidal functions of rabbit alveolar and peritoneal macrophages: effects of non-specific activation and a natural microbicidal peptide, MCP-1, *Aust. J. Exp. Biol. Med. Sci.*, 63, 53, 1985.

26. **Flesch, I. E. A. and Kaufmann, S. H. E.**, Activation of tuberculostatic macrophage functions by gamma-interferon, IL-4, and tumor necrosis factor, *Infect. Immunol.*, 58, 2675, 1990.

27. **Kos, F. J.**, Augmentation of recombinant interleukin-2-dependent murine macrophage mediated tumor cytotoxicity by recombinant tumor necrosis factor-alpha, *Immunol. Cell Biol.*, 67, 433, 1989.

28. **Frendl, G. and Beller, D. I.**, Regulation of macrophage activation by IL-3. I. IL-3 functions as a macrophage-activating factor with unique properties, including Ia and lymphocyte function-associated antigen-1, but not cytotoxicity, *J. Immunol.*, 144, 3392, 1990.

29. **Frendl, G. and Beller, D. I.**, Regulation of macrophage by interleukin-3. Interleukin-3 and lipopolysaccharide act synergistically in the regulation of interleukin-1 expression, *J. Immunol.*, 144, 3400, 1990.

30. **Phillips, W. A., Croatto, M., and Hamilton, J. A.**, Priming the macrophage respiratory burst with interleukin-4: enhancement with TNF-alpha but inhibition by IFN-gamma, *Immunology*, 70, 498, 1990.

31. **Elliot, G. R., Van Batenburg, M. J., and Dzolic, M. R.**, Morphine dependence modified macrophage eicosanoid release, *Agents Actions*, 23, 32, 1987.

32. **Vogel, S. N., Marshall, S. T., and Rosenstreich, D. L.**, Analysis of the effects of lipopolysaccharide on macrophages: differential phagocytic responses of C3H/HeN and C3H/HeJ macrophages *in vitro*, *Infect. Immun.*, 25, 328, 1979.

33. **Glode, L. M., Jaques, A., Mergenhagen, S. E., and Rosenstreich, D. L.**, Resistance of macrophages from C3H/HeJ mice to the *in vitro* cytotoxic effects of endotoxin, *J. Immunol.*, 119, 162, 1977.

34. **Leslie, C. C., McCormick-Shannon, K., Cook, J. L., and Mason, R. J.**, Macrophages stimulate DNA synthesis in rat alveolar type II cells, *Am. Rev. Respir. Dis.*, 132, 1246, 1985.

35. **Stadnyk, A. W., Befus, A. D., and Gauldie, J.**, Characterization of nonspecific esterase activity in macrophages and intestinal epithelium of the rat, *J. Histochem. Cytochem.*, 381, 1, 1990.

36. **Collier, H. O. J., Francis, D. L., Mcdonald-Gibson, W. J., Roy, A.C., and Saeed, S. A.**, Prostaglandins, cyclic AMP and the mechanism of opiate dependence, *Life Sci.*, 17, 85, 1975.

37. **Casale, T. B. and Kaliner, M.**, A rapid method for isolation of human mononuclear cells free of significant platelet contamination, *J. Immunol. Methods*, 55, 347, 1982.

38. **Skaug, V., Davies, R., and Gylseth, B.**, *In vitro* macrophage cytotoxicity of five calcium silicates, *Br. J. Indust. Med.*, 41, 116, 1984.

39. **Vogel, S. N., Moore, R. N., Sipe, J. D., and Rosenstreich, D. C.**, BCG-induced enhancement of endotoxin sensitivity in C3H/HeJ mice. I. *In vivo* studies, *J. Immunol.*, 124, 2004, 1980.

40. **Stuehr, D. J. and Marletta, M. A.**, Mammalian nitrate biosynthesis: mouse macrophages produce nitrite and nitrate in response to *Escherichia coli* lipopolysaccharide, *Proc. Natl. Acad. Sci. U.S.A.*, 82, 7738, 1985.

41. **Miwa, M., Stuehr, D. J., Marletta, M. A., Wishnok, J. S., and Tannenbaum, S. R.**, Nitrosation of amines by stimulated macrophages, *Carcinogenesis*, 8, 955, 1987.

42. **Bjerknes, R., Bassoc, C., Sjursen, H., Laerum, O., and Solberg, C.**, Flow cytometry for the study of phagocyte function, *Rev. Infect. Dis.*, 2, 16, 1989.

43. **Doolittle, M., Bohman, R., Durstenfeld, A., and Cascarano, J.**, Identification and characterization of liver nonparenchymal cells by flow cytometry, *Hepatology*, 7, 696, 1987.

44. **Steinkamp, J. A., Wilson, J. S., Saunders, G. C., and Stewart, C. C.**, Phagocytosis: flow cytometric quantitation with fluorescent microspheres, *Science*, 215, 64, 1982.
45. **Stewart, C. C., Lehnert, B. E., and Steinkamp, J. A.**, *In vitro* and *in vivo* measurement of phagocytosis by flow cytometry, *Methods Enzymol.*, 132, 183, 1986.
46. **Darzynkiewicz, Z. and Andreff, M.**, Multiparameter flow cytometry. I. Application in analysis of the cell cycle, *Clin. Bull.*, 11, 47, 1981.
47. **Pollack, A., Prudhomme, D., Greenstein, D., Irvin, G., Clafin A., and Block, N.**, Flow cytometric analysis of RNA content in different cell populations using pyronin Y and methyl green, *Cytometry*, 3, 28, 1982.
48. **Braylan, R. C., Benson, N. A., Nourse, V., and Kruth, H. S.**, Correlated analysis of cellular DNA, membrane antigens and light scatter of human lymphoid cells, *Cytometry*, 2, 337, 1982.
49. **Koren, H. S.**, Antibody-dependent cellular cytotoxicity — lysis of hapten-modified target cells, in *Manual of Macrophage Methodology*, Herscowitz H. B., Holden, H. T., Bellanti, J. A., and Ghaffar, A., Eds., Marcel Dekker, New York, 1981, 315.
50. **Rosenberg, E. B., McCoy, J. L., Green, S. S., Donelly, F. C., Siwarski, D. F., Levine, P. H., and Herberman, R. B.**, Destruction of human lymphoid tissue-culture cell lines by human peripheral lymphocytes in[51] Cr-release cellular cytotoxic assays, *J. Natl. Cancer Inst.*, 52, 345, 1974.
51. **Donahoe, R. M.**, Commentary on: in Vitro immunotoxicology and immunopharmacology: studies on drugs of abuse, *Toxicol. Lett.*, 53, 265, 1990.
52. **Lohr, K. M. and Snyderman, R.**, *In vitro* methods for the study of macrophage chemotaxis, in *Manual of Macrophage Methodology*, Herscowitz, H. B., Holden, H. T., Bellanti, J. A., and Ghaffar, A., Eds., Marcel Dekker, New York, 1981, 303.
53. **Park, B. H., Dolen, J., and Snyder, B.**, Defective chemotactic migration of PMNs in patients with Pelger-Huet anomaly, *Proc. Soc. Exp. Biol. Med.*, 155, 51, 1977.
54. **Lee, T. P., Moscati, R., and Park, B.**, Effects of pesticide on human leukocyte function, *Res. Comm. Chem. Pathol. Pharmacol.*, 23, 597, 1979.
55. **Zurier, R., Weissman, G., Hoffstein, S., Kammerman, S., and Tai, H.**, Mechanism of lysosomal enzyme release from human leukocytes, *J. Clin. Invest.*, 53, 297, 1974.
56. **Tan, J., Watanakunakorn, C., and Phair, J.**, A modified assay of neutrophil function: use of lysostaphin to differentiate defective phagocytosis from impaired intracellular killing, *J. Lab. Clin. Med.*, 78, 316, 1971.
57. **Ferrante, A. and Thong, Y. H.**, A rapid one-step procedure for purification of mononuclear and polymorphonuclear leukocytes from human blood using a modification of the Hypaque-Ficoll technique, *J. Immunol. Methods*, 24, 384, 1978.
58. **Boyum, A.**, A one-stage procedure for isolation of granulocytes and lymphocytes from human blood, *Scand. J. Clin. Lab. Invest.*, 21, 51, 1968.
59. **Matthews, N. and MacLaurin, P. B.**, Spontaneous cytotoxic activity as a test of human lymphocyte function, *Lancet*, 1, 581, 1975.
60. **Bushkin, S. C., Pantic, V. S., and Good, R. A.**, RNA and protein synthesis in spontaneous rosette formation by T-lymphocytes, *J. Immunol.*, 115, 866, 1975.
61. **Dolen, J. G. and Park, B. H.**, An improved micromethod for enumeration of human B cell rosettes with mouse red blood cells, *Immunol. Commun.*, 7, 677, 1978.
62. **Sorg, C.**, Lymphokines, monokines, and cytokines, *Chem. Immunol.*, 49, 82, 1990.
63. **Morgan, D. A., Ruscetti, F. W., and Gallo, R.C.**, Selective *in vitro* growth of T-lymphocytes from normal human bone marrow, *Science*, 193, 1007, 1976.
64. **Hirschorn, R.**, The effect of exogenous nucleotides on the response of lymphocytes to PHA, PWM and Con A, in *Cyclic AMP, Cell Growth and the Immune response*, Braun, W., Lichtenstin, M., and Parker, C. W., Eds., Springer-Verlag, New York, 1974, p. 45.
65. **Campbell, A. K.**, Lymphocyte activation, in *Intracellular Calcium*, Campbell, A. K., Ed., John Wiley & Sons, New York, 1983, 376.

66. **Safe, S.**, PCBs and PBBs: biochemistry, toxicology and mechanism of action, *Crit. Rev. Toxicol.*, 13, 319, 1984.

67. **Petersen, B. H., Graham, J., and Lemberger, L.**, Marijuana, tetrahydrocannabinol and T cell function, *Life Sci.*, 19, 395, 1976.

68. **Puess, M. M. and Lefkowitz, S. S.**, Influence of maturity on immunosuppression by delta-9 tetrahydrocannabinol, *Proc. Soc. Exp. Biol. Med.*, 158, 350, 1978.

69. **White, S. C., Brin, S. C., and Janicki, B. W.**, Mitogen-induced blastogenic responses of lymphocytes from marijuana smokers, *Science*, 188, 71, 1975.

70. **Wybran, J., Appelboom, T., Famey, J. P., and Govarto, A.**, Suggestive evidence for receptors for morphine and methionine enkephalinon normal human blood lymphocytes, *J. Immunol.*, 123, 1068, 1979.

71. **Klein, T. W., Heston, C. A., and Friedman, H.**, Suppression of human and mouse lymphocyte proliferation by cocaine, *Adv. Biochem. Psychopharmacol.*, 44, 139, 1988.

72. **Wojdani, A. L., Attarzadeh, M., Wolde-Tsadik, G., and Alfred, L. A.**, Immunocytotoxicity effects of polycyclic aromatic hydrocarbons on mouse lymphocytes, *Toxicology*, 31, 181, 1984.

73. **Klempau, A. E. and Cooper, E. L.**, T-Lymphocyte and B-lymphocyte dichotomy in anuran amphibians. I. T-lymphocyte proportions, distribution and ontogeny, as measured by E-rosetting, nylon wool adherences, postmetamorphic thymectomy, and nonspecific esterase staining, *Devel. Comp. Immunol.*, 7, 99, 1983.

74. **Boyum, A., Berg, T., and Blomhoff, R.**, Fractionation of mammalian cells, in *Iodinated Density Gradient Media: A Practical Approach*, Rickwood, D., Ed., IRL Press, Oxford, 1983, 147.

75. **Alfred, L. and Wojdani, A.**, Effects of methylcholanthrene and benzanthracene on blastogenesis and aryl hydrocarbon hydroxylase induction in splenic lymphocytes from three inbred strains of mice, *Int. J. Immunopharmacol.*, 5, 123, 1983.

76. **Luster, M. I., Portier, C., Pait, D. G., White, K. L., Jr., Gennings, C., Munson, A. E., and Rosenthal, G. J.**, Risk assessment in immunotoxicology. I. Sensitivity and predictability of immune tests, *Fund. Appl. Toxicol.*, 18, 200, 1992.

77. **Luster, M. I., Munson, A. E., Thomas, P. T. Holsapple, M. P., Fenters, J. D., White, K. L. Jr., Lauer, L. D., Germolee, D.R., Rosenthal, G. J., and Dean, J. H.**, Development of a testing battery to assess chemical-induced immunotoxicity: National Toxicology Program's guidelines for immunotoxicity evaluation in mice, *Fundam. Appl. Toxicol.*, 10, 2, 1988.

78. **Dean, J. H., Cornacoff, J. B., Rosenthal, G. J., and Luster, M. I.**, Immune system: Evaluation of injury, in *Principles and Methods of Toxicology*, 2nd ed., Hayes, A. W., Ed., Raven Press, New York, 1989, 741.

79. **Trizio, D., Basketter, D. A., Botham, P. A., Graepel, P. H., Lambre, C., Magda, S. J., Pal, T. M., Riley, A. J., Ronneberger, H., Van Sittert, N. J., and Bontinck, W. J.**, Identification of immunotoxic effects of chemicals and assessment of their relevance to man, *Food. Chem. Toxicol.*, 26, 527, 1988.

80. **Wick, G.**, A bridgehead for alternative methods: immunology, in *Scientific Alternatives to Animal Experiments*, Lembeck, F., Ed., Ellis Horwood, Chichester, Great Britain, 1989, 153.

Chapter 8

PHARMACOKINETIC STUDIES IN CELLULAR SYSTEMS

I. INTRODUCTION

Pharmacokinetics, or *toxicokinetics*, is the knowledge of what happens to a dose of a drug after it is administered and how this dose is related to its concentration in the target organ. The *target organ* may be defined as the tissue or cells to which the drug distributes. Toxicokinetics includes knowledge about the absorption of the toxic dose of a chemical in the gastrointestinal tract, through the skin, via the lungs, or across the mucous membranes. Once a percentage of the dose is absorbed, it may be chemically modified into a more or less toxic compound. This metabolism occurs primarily in the liver, but may also take place in the lungs or kidneys. Whether or not the chemical is metabolized, it is distributed to various organs or it may remain in the blood, depending on its physicochemical properties. Finally, the substance is eliminated either actively or passively through the action of the kidneys, lungs, skin, or the intestinal tract.

The toxicologist makes an estimation of the risk of the chemical to humans based on the effects of different doses on animals and whether this method of exposure is analogous to that in the human. When using cell cultures, it is only possible to obtain information about the effects of different concentrations in the target organ. These efforts must therefore be supplemented with studies ready to determine the pharmacokinetic disposition of a chemical. This information is important if toxicity testing in cell cultures is to be used for risk assessment. Not surprisingly, most pharmacokinetic effects can be measured in cell systems as relatively simple interactions between chemical substances and cell functions.

This chapter will introduce the concept of kinetic distribution and metabolism of test compounds in *in vitro* systems and will coordinate the information currently available with future studies.

II. METABOLIC STUDIES IN CELL CULTURE

The liver plays a key role in the detoxification of foreign substances, but it can also convert the substance into a toxic metabolite. It is possible for the toxic metabolite to induce damage to the liver or other target organs. The underlying mechanism for liver injury may be explained through an organ-specific effect.

In primary cultures of hepatocytes, it is possible to study both metabolite-induced liver toxicity and the pharmacological aspect of liver metabolism, i.e., the detoxification or activation of chemicals. Cells from chicken, mouse, rat, and man have been used and maintained in culture, either as organ cultures,

121

monolayers, or as isolated cells in suspension. The cells from mouse or rat are obtained by perfusion of the liver using collagenase to separate the cells rapidly and gently. Cytochrome P450 enzymes are functional in these systems from a few hours to several days. Cultures of human hepatocytes have also been realized with the coculture of feeder layers of other continuous cell lines. Recently, functional human hepatocytes have been maintained in a viable state after freezing.[1] This is valuable because liver metabolism is species specific. Another development is the use of cell lines, with preserved P450 activity, instead of primary cultures of liver cells. Finally, several systems coculturing hepatocytes and other target cells have been used for pharmacokinetic studies.[2-4]

Among the different types of cultures just described, metabolic and hepatocellular transporting properties are studied with a large number of substances. Examples of specific research areas include (1) comparison between the metabolizing capacity of liver cells and S9 mixtures; (2) how cytochrome P450 activity may be preserved in the hepatocytes; (3) species-specific metabolism; (4) the protective effects of reduced glutathione (GSH) in the presence of liver damage; and, (5) comparison between the metabolizing capacity of the liver and that of other organs, such as the kidney, lung, and intestine.[5]

III. *IN VITRO* STUDIES OF THE ABSORPTION, DISTRIBUTION, AND ELIMINATION OF CHEMICALS

The uptake and distribution of a chemical in the body, as well as its elimination, are determined by biotransformation and other processes, such as (1) the ability and time that a substance requires to cross cellular layers, including intestinal mucous membranes and the blood-brain barrier; (2) the ability of the substance to enter and bind to different cells; (3) the extent of binding to blood proteins and stored fat; (4) the passage through the capillary endothelium; and (5) the ability of the substance to disperse in the water phase of the intestine, blood, extracellular fluid, and intracellular fluid, as a function of its water-soluble properties.

It is possible to predict the reactivity and the biotransformation of a chemical from its structural formula. It is also possible to predict the uptake, distribution, and elimination from the structure of the substance and whether it activates certain specific cellular transport mechanisms. The physicochemical properties of the chemical, such as water and lipid solubility, ionization constant in acid or basic pH, and protein binding capacity, can also be used to predict its pharmacokinetic characteristics. For example, large and highly ionized molecules have difficulty in entering or passing through cells and will therefore be poorly absorbed or will not be able to penetrate the blood-brain barrier. Small and moderately acidic or basic molecules, on the other hand, are absorbed and distributed to the extra- and intracellular spaces. Finally, highly lipophilic substances are readily absorbed and rapidly distributed to all compartments. In addition, they are often bound to tissue in high concentrations, and this may also account for the high degree of intracellular protein binding. For given doses of the three types of molecules it is possible to predict their concentration in the tissue and their ability to enter the brain (and their potential for neurotoxicity).

Most mutagenic and carcinogenic substances can affect a cell at relatively low doses. This is accounted for by biotransformation of the substance, and therefore is responsible for classifying most mutagens as indirect acting. Substances having general toxic effects, especially after acute toxic doses, demonstrate toxicity at higher concentrations which saturate the biotransforming capacity of the liver. This concentration can be estimated if the dose and the physicochemical properties of the substance are known. *In vitro* tests to determine the distribution coefficient, protein binding, dissociation, and cellular transport are now easily performed. The physicochemical data are used to supplement other types of toxicity data.

IV. BIOTRANSFORMATION OF CHEMICALS IN CELL CULTURE

A. ENZYMATIC METABOLISM

Most cells in the body have the capacity to enzymatically metabolize chemicals to less toxic and more water-soluble compounds, which later may be excreted through the urine. This capacity is clearly most pronounced in liver cells, but kidney, lung, intestine, and skin have extensive capacity for xenobiotic metabolism.[6] Most other organs, such as blood cells, endothelial cells, muscle, and connective tissue, have a relatively low capacity for biotransformation. The main aspect of the biotransformation of chemicals by the most active cells is the ability to add functional groups (phase I reactions), which later may be bound to endogenous substances, such as glucuronic acid or glutathione, resulting in increased water solubility of the metabolite (phase II biotransformation). The most important phase I enzyme is cytochrome P450 oxygenase, inducible by such chemicals as phenobarbital. Examples of other, more commonly found enzymes are epoxide hydrolases (phase I) and the transferases (phase II). It is important to note that the described biotransformation in some cases may result in more reactive (toxic) metabolites. Details of these reactions are discussed in Chapter 4, Section I, and in Chapter 9.

Maintenance cultures of specific cells with high metabolic activity display biotransformation capacity, including significant P450 activity, for several hours or days. There is some question if this short time period is sufficient to mimic clinically relevant metabolism of all types of chemicals. Early passages of finite lines, especially of fetal cells, may display certain metabolic activity, whereas most other cultures of both finite and continuous cell lines demonstrate low or immeasurable P450 levels, which sometimes may be coupled to hydrolase or glucuronyl transferase activity. The P450 activity found in some cell lines can be increased by induction *in vitro*. Thus, few cell lines have some biotransformation capacity as the least competent cell types *in vivo*.

B. THE S9 MIXTURE

Presently, there are two methods available to increase the capacity of biotransformation of cell cultures: (1) the target cells are cultured on a feeder layer of more metabolically competent cells, e.g., BHK21 cells irradiated to inhibit mitotic division; (2) there may be added to the culture medium whole microsomes, or preferably an ultracentrifuged fraction of homogenized liver from rats or mice which have been pretreated with Arochlor 1254, thus containing P450 and other important microsomal and cytosol enzymes. This fraction plus the cofactor NADPH is called the S9 mixture.[7] (See Figure 1.)

S9-reinforced cell line cultures offer moderate advantages over primary cultures of hepatocytes:

1. The system is flexible, technically easy to prepare, and can be used with a variety of cell culture models.
2. Drug-metabolizing enzymes are relatively stable in storage and can be characterized biochemically prior to incorporation into the experiment.
3. The system enhances phase I reactions, such as the formation of reactive metabolites, but induces relative decreases of the biosynthetic phase II reactions, since the energy-rich phase II cofactors are not included in the mix.

More significant disadvantages often discourage the use of the mix:

1. S9-supplemented cultures require the addition of exogenous cofactors, such as NADPH.
2. It is cytotoxic to many systems, and test results and the enzyme activity, per milligram of protein, varies considerably.

A thorough discussion of the history and uses of the S9 system is found in Chapter 9.

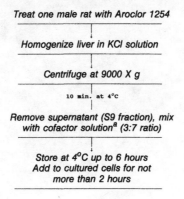

Figure 1. Protocol for S9 mixture. [a] MgCl, glucose-6-phosphate, NADPH in phosphate butter (0.2 mol/l pH 7.4). (Modified from Flint, O. P. and Orton, T. C., *Toxicol. Appl. Pharmacol.*, 76, 383, 1984.)

REFERENCES

1. **Gomez-Lechon, M. J., Lopez, P., and Castell, J. V.**, Biochemical functionality and recovery of hepatocytes after deep freezing storage, *In Vitro*, 20, 826, 1984.
2. **Aszalos, A., Bradlaw, J. A., Reynaldo, E. F., Yang, G. C., and El-hage, A. N.**, Studies on the action of nystatin on cultured rat myocardial cells and cell membranes, isolated rat hearts and intact rats, *Biochem. Pharmacol.*, 33, 3779, 1984.
3. **Bandiera, S., Sawyer, T. W., Campbell, M. A., Fujita, T., and Safe, S.**, Competitive binding to the cytosolic 2,3,7,8-tetrachlorodibenzo-*p*-dioxin receptor, *Biochem. Pharmacol.*, 32, 3803, 1983.
4. **Bandiera, S., Sawyer, T., Ramkes, M., Zmudzka, B., Safe, L., Mason, G., Keys, B., and Safe, S.**, Polychlorinated dibenzofurans (PCDFs): effects of structure on binding to the 2,3,7,8-TCDD cytosolic receptor protein, AHH induction, and toxicity, *Toxicology* 32, 131, 1984.
5. **Gram, T. E., Okine, L. K., and Gram, R. A.**, The metabolism of xenobiotics by certain extrahepatic organs and its relation to toxicity, *Annu. Rev. Pharmacol. Toxicol.*, 26, 259, 1986.
6. **Gram, T. E.**, The pulmonary mixed-function oxidase system, in *Toxicology of Inhaled Materials*, Witschi, H. P. and Brain, J. D., Eds., Springer-Verlag, Berlin, 1985, 421.
7. **Flint, O. P. and Orton, T. C.**, An *in vitro* assay for teratogens with cultures of rat embryo midbrain and limb bud cells, *Toxicol. Appl. Pharmacol.*, 76, 383, 1984.

Chapter 9

CELLULAR METHODS OF GENOTOXICITY AND CARCINOGENICITY

Thomas W. Sawyer

I. INTRODUCTION

It has long been recognized that it is not feasible to perform extensive animal testing on the thousands of chemicals which are present in society and the environment. Nevertheless, the potential hazards that these chemicals pose require that some effort be made to obtain information on them in order to prioritize further studies. This has led to the development of a wide variety of methods that are quicker, more economical, and more convenient than whole animal tests to detect mutations, chromosome breakage, and other genetic effects. Until recently, these short-term tests were used primarily to detect chemical mutagens as indicators of their potential carcinogenic activity. As our understanding of the complex processes leading to cancer has improved, however, short-term tests have also been developed to reflect these advancements.

A. HISTORY

The basis for the short-term testing approach is derived from observations and carcinogenesis studies that indicate that the induction of many cancers can be traced to chemical exposures which cause mutational events (Table 1). In the late 18th century, Sir Percival Pott, an English physician, noted that many of his patients who had scrotal cancer were chimney sweeps, and hypothesized that soot and coal tars were the causative agents.[1] During the following years, several additional associations between mostly occupational chemical exposures and cancer induction were also made. It was not until 1915, however, that cancer was first chemically induced in an animal, when Yamagiwa and Ichikawa discovered that repeated applications of coal tar to rabbit ears produced epidermal tumors.[2,3] Kennaway and co-workers, in exhaustive studies that involved fractionating two tons of gasworks pitch, isolated the active ingredients and identified them as polycyclic aromatic hydrocarbons (PAH) consisting mainly of benzo[a]pyrene (B[a]P).[4] In 1947, Boyland first suggested that these relatively inert substances required metabolism in order to exert their carcinogenic activity,[5] and subsequent work by Miller and Miller[6] extended this concept to demonstrate that a large proportion of carcinogens in several chemical classes fell into this category.

TABLE 1
Genesis of Short-Term Tests for Potential
Genotoxic Carcinogenicity

Date	Test	Ref.
1775	Soot and coal tar associated with human cancer	Pott, 1775 Ref. 1
1914	Somatic cell mutation suggested to be the first stage of cancer	Boveri, 1914 Ref. 7
1915	Chemical induction of epidermal tumors on ears of rabbits treated with coal tar	Yamagiwa and Ichikawa, 1915 Ref. 2, 3
1927	Induction of mutations in *Drosophila* with X-rays	Muller, 1927 Ref. 8
1932	Isolation and identification of benzo[*a*]pyrene from tar as the causative agent in the induction of cancer	Cook et al., 1932 Ref. 4
1946	Chemical induction of mutations in *Drosophila* by sulfur mustard	Auerbach and Robson, 1946 Ref. 9
1947	Proposed that polycyclic aromatic hydrocarbons require metabolism for carcinogenic activity	Boyland, 1947 Ref. 5
1964	Correlation demonstrated between degree of binding to DNA and carcinogenic potency of five polycyclic aromatic hydrocarbons	Brookes and Lawley, 1964, Ref. 10
1969	Proposal that most classes of chemical carcinogens require metabolism before they become active	Miller and Miller, 1969, Ref. 6
1971	Activation of dimethylnitrosamine to a bacterial mutagen, using a liver homogenate	Malling, 1971 Ref. 11
1973	Tests developed to detect carcinogens as mutagens using a *Salmonella*/S9 system	Ames et al., 1973 Ref. 12
1975	High success rates demonstrated in predicting carcinogens as mutagens in the Ames test	Ames et al., 1975 Ref. 13, 14
Post 1975	General acceptance of some short-term tests as useful indications of potential carcinogens, but also, realization that most short-term tests are not appropriate to detect all classes of chemical carcinogens	

1. Benzo[a]pyrene

Figure 1 illustrates the metabolic activation of the potent carcinogen B[a]P. B[a]P itself is not carcinogenic, but requires metabolic activation. As an inactive "procarcinogen", B[a]P is first enzymatically catalyzed to the intermediate B[a]P-7,8-epoxide and then to B[a]P-7,8-diol. These intermediates are referred to as *proximate carcinogens*, since they are more carcinogenic than the parent molecule, but still require further metabolism to exert carcinogenic activity. B[a]P-7,8-diol is then further activated to the ultimate carcinogen, B[a]P-7,8-diol-9,10-epoxide. The ultimate, or reactive, form of a carcinogen is electrophilic and capable of interacting with electron-rich cellular macromolecules, like DNA. Although the pathway represented here leads to the most carcinogenic form of B[a]P, it should be noted that it is simplified and only one of many competing pathways in the complex metabolism of B[a]P.

Concurrent with the early carcinogenesis studies were rapid advancements in the field of genetics. In 1914 Boveri proposed that cancer was the result of somatic cell mutation,[7] and by 1927 the modern field of mutation research was born when Müller induced mutations in *Drosophila* with X-rays.[8] Auerbach and Robson first discovered the existence of chemically induced mutations in 1942, when they found that mustard gas (bis 2-chloroethyl sulfide) induced mutations in *Drosophila*, but only published their findings in 1946[9] when wartime censorship was relaxed. The chemical mutagenesis and carcinogenesis fields effectively merged when Brookes and Lawley showed that the carcinogenic potency of five PAHs correlated with their binding affinities to DNA, but not RNA or proteins.[10] Based on this correlation, short-term tests with bacteria were developed to detect known carcinogens. Although these tests were sensitive to direct-acting carcinogens, i.e., chemicals active without prior metabolism, most carcinogens tested were inactive. The breakthrough came when Malling adopted the suggestion of Miller and Miller and others, suggesting that most of the enzymes responsible for xenobiotic metabolism could be found in the endoplasmic reticulum of liver cells. By incorporating a crude homogenate of liver he was able to activate dimethylnitrosamine to a chemical moiety which was mutagenic for bacteria.[11] Bruce Ames also adopted this approach and in a series of landmark papers showed that bacterial tests for mutagenesis were highly predictive of potential chemical carcinogenesis.[12–14]

While the introduction of the Ames test may well be viewed as the birth of the modern era of short-term testing, the initial promise of this type of approach never quite materialized. The fact that the process of carcinogenesis is far more complex than originally foreseen is part of the explanation. For this reason, assays based on only one mutational event (or any single event) could not possibly provide an adequate screen for chemical carcinogens acting by a number of distinct and overlapping mechanisms. Although it is beyond the scope of this chapter to discuss the different types of chemical carcinogens in detail (for a good overview, see Reference 15) it is instructive to note that they can be broadly divided into three groups as shown in Table 2.[15] The first group

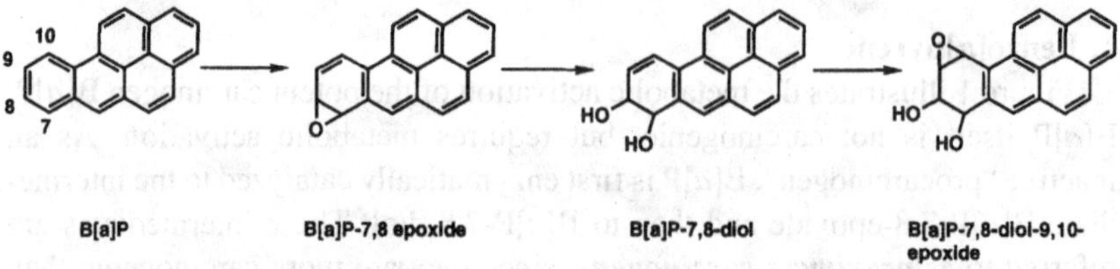

Figure 1. Simplified schematic of the metabolic activation of benzo[*a*]pyrene (B[*a*]P) to its ultimate carcinogenic form. B[*a*]P is enzymatically converted to the proximate carcinogens B[*a*]P-7,8-epoxide and B[*a*]P-7,8-diol, and then to the ultimate carcinogen, B[*a*]P-7,8-diol-9,10-epoxide.

contains chemicals that apparently exert their carcinogenic effects at the level of DNA and includes either direct-acting or indirect-acting chemicals. The second group of compounds is known not to interact with DNA, i.e., they are "epigenetic" carcinogens, and have also been identified with known biological effects that could lead to cancer. Examples of this group include tumor promoters, peroxisome proliferators, immunosuppressors, solid state carcinogens, and agents that are cytotoxic or modify hormones. The last group is comprised of compounds which do not react with DNA, but whose mechanism of action has not been elucidated.

2. The Polycyclic Aromatic Hydrocarbons

The PAHs are the class of carcinogens whose mechanism of action is probably the most well understood. They represent one of the most significant human cancer hazards and are the group of compounds most often assessed for carcinogenicity in rodents and in short-term testing protocols. A description of the steps that a prototypical PAH like B[*a*]P might take as it causes a malignant transformation in target cells (Figures 2 and 3) illustrates the complexity of the carcinogenic process, as well as some of the endpoints used as the basis of a number of short-term testing protocols.

As mentioned above, the activation of B[*a*]P to its ultimate form is catalyzed by the drug-metabolizing or biotransformation enzymes (for review, see Reference 16). These enzyme systems have been the subject of intense study that has resulted in innumerable publications on their characterization and effects. Localized primarily in the liver, they are capable of acting on a wide variety of structurally diverse chemical classes and are classified as possessing either phase I or phase II type reactivity. Phase I enzymes catalyze oxidative, reductive, and hydrolytic reactions that generally increase the water solubility of their substrates, but also act to provide a site for the catalytic activity of the phase II enzymes which are conjugative or synthetic in nature. The most important phase I enzymes are the microsomal cytochrome P450-containing enzymes which are divided into several different families, depending on their polypeptide structure and catalytic activity. Several of these families are characteristically induced by specific chemicals which can elevate their levels dramatically. The phase I and II profiles of enzymes act in concert, and it is the

TABLE 2
Classification of Carcinogenic Chemicals

Category and class	Example
A. DNA-reactive (genotoxic) carcinogens	
1. Activation-independent organic	Alkylating agents
2. Activation-independent inorganic	Nickel, cadmium
3. Activation-dependent	Polycyclic aromatic hydrocarbon, arylamine, nitrosamine
B. Epigenetic carcinogens	
1. Promoter	Organochlorine pesticides, saccharin
2. Hormone-modifying	Estrogen, amitrol
3. Peroxisome proliferators	Chlofibrate, diethylhexylphthalate
4. Cytotoxic	Nitrilotriacetic acid
5. Immunosuppressor	Cyclosporin A, azathioprine
6. Solid state	Plastics, asbestos
C. Unclassified	
1. Miscellaneous	Ethanol, dioxane

From Williams, G.M. and Weisburger, J.H., in *Casarett and Doull's Toxicology: The Basic Science of Poisons, 4th ed.*, Amdur, M.O. et al., Eds., Pergamon Press, New York, 1991, 127. With permission.

equilibrium attained between these enzymes that determines just how much a procarcinogen like B[a]P is metabolized to its ultimate carcinogenic form or to innocuous products. Probably the major shortcoming of most short-term tests used today is the lack of a complete complement of the biotransformation enzymes.

Once B[a]P has been activated, it must then bind to critical macromolecules in the target cell, for example, the DNA of basal epidermal cells in the skin. Covalent binding of the carcinogen to DNA is a necessary first step in carcinogenesis and several studies have extended Brookes and Lawley's pioneering work[10] to show that the extent of covalent binding to DNA correlates well with a compound's carcinogenic activity in mouse skin[17,18] and liver.[19,20] DNA binding is a direct measure of a chemical's potential genotoxicity and carcinogenicity. Until recently, however, most studies that examined DNA binding and adduct formation utilized radiolabeled compounds and high pressure liquid chromatography,[21] thus limiting work to chemicals that could be radiolabeled. A recent advance has been the introduction of postlabeling techniques, where the adducts are labeled with a radioactive isotope after they have been formed. The most sensitive method involves digesting the DNA samples enzymatically and then labeling the adducts with ^{32}P.[22-25] The adducts are then resolved and identified chromatographically against prepared standards. This approach appears to have great potential for short-term screening as evidenced

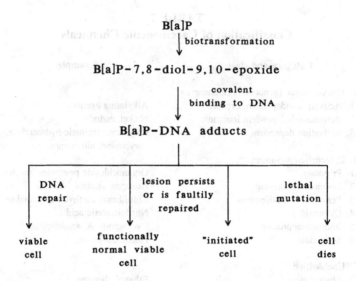

Figure 2. Some possible fates of cells containing B[*a*]P-DNA adducts.

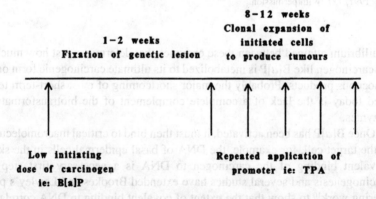

Figure 3. Initiation-promotion model of chemical carcinogenesis. The target tissue, i.e., skin, is treated with low "initiating" doses of a carcinogen like benzo[*a*]pyrene (B[*a*]P). The genetic lesions caused by the covalent binding of these compounds to DNA, are then allowed to become fixed into the genome by cellular DNA replication for 1 to 2 weeks. These initiated cells remain dormant unless promoted to become tumors by repeated application of a promoter like 12–*O*-decanoyl-phorbol-13-acetate (TPA).

by studies where aromatic adducts were detected in the placentas of mothers who smoked during pregnancy.[26]

Once the active form of a carcinogen is bound to DNA, several paths can be taken (Figure 2): (1) the adducts are removed and the lesion is repaired by enzymatic DNA repair processes; (2) the lesion persists or is incompletely repaired, resulting in a functionally normal viable cell; (3) the lesion persists or is improperly repaired, resulting in neoplastic conversion of the cell; or (4) the cell dies as a result of lethal mutations. It is at this point in the theoretical development of cancer that most short-term tests have based their end points. As a consequence, most short-term tests are designed to detect mutations, DNA repair processes, or chromosome damage.

3. Initiation, Promotion, and Progression

Cancer develops in stages in most tissues, and the cell that is neoplastically converted takes a path towards malignancy whose direction is dependent both on the type and dose of carcinogen, as well as chemical exposure after the fact. The original two-stage initiation-promotion model, shown in Figure 3, has been modified to include a third stage so that the sequential development of cancer, from a normal cell to a fully malignant tumor, is now thought to include at least three distinct stages termed *initiation, promotion,* and *progression*.[27]

In experimental animal models, *initiation* of the target cell population is caused by the administration of single, small "initiating" doses of carcinogen. The binding of the carcinogen to the DNA of the cell is an irreversible event whose effectiveness is dependent upon how soon this lesion is "fixed" by cellular replicative DNA synthesis. The second stage, tumor *promotion,* is primarily distinguished from the stages of initiation and progression by its reversibility. If the initiated cell population is repeatedly treated with a promoter such as 12-*O*-decanoyl-phorbol-13-acetate (TPA), then a proportion of this population will be stimulated to proliferate, resulting in the formation of predominantly benign tumors. Cessation of TPA treatment results in the majority of these tumors regressing. The *progression* stage of carcinogenesis is characterized by irreversible genetic changes that lead from the development of benign tumors to those having malignant properties. Tumors are also induced in animals by complete carcinogenesis protocols, where a single large dose or repeated applications of smaller doses of a carcinogen are administered to the animal. In this case, all the components of the multistage concept of carcinogenesis are presumed to be present. With the exception of cell transformation assays, there are very few short-term tests that reflect the multistage nature of chemical carcinogenesis, or that can detect promoters. This is a drawback when screening for potential carcinogens using short-term tests, especially when it is now clear that exposure to many chemicals can enhance cancer risk through mechanisms other than a genetic "initiating" event.

B. VALIDATION AND PREDICTIVE ABILITY OF SHORT-TERM TESTS

The field of short-term testing has not been without controversy. Ames' initial success at detecting carcinogens raised a great deal of enthusiasm for this approach and for several years his assertion that "mutagens are carcinogens" was corroborated as other laboratories repeatedly obtained high success rates

TABLE 3
Definition of Terms Describing Validation Study Results

Carcinogenicity classification	Test result +	−
Carcinogen	a	b
Noncarcinogen	c	d

$$\text{Sensitivity} = \frac{\text{Number of carcinogens positive in test}}{\text{Total number of carcinogens}} = \frac{a}{a+b}$$

$$\text{Specificity} = \frac{\text{Number of non carcinogens negative in test}}{\text{Total number of carcinogens}} = \frac{d}{c+d}$$

$$\text{Predictive value} = \frac{\text{Number of positive results from carcinogens}}{\text{Total number of positive results}} = \frac{a}{a+c}$$

$$\text{Concordance} = \frac{\text{Number of correct predictions}}{\text{Total number of chemicals}} = \frac{a+d}{a+b+c+d}$$

$$\text{Prevalance} = \frac{\text{Number of carcinogens}}{\text{Total number of chemicals}} = \frac{a+b}{a+b+c+d}$$

Adapted from Purchase, I. F. H., *Mutat. Res.*, 99, 53, 1982. With permission.

when screening what turned out to be select groups of compounds. As shown above, however, the reasoning behind the Ames test was fatally flawed when applied across several classes of carcinogens, and this soon became apparent as attempts to validate these tests became more rigorous.

The validation of predictive tests is vital to determine just how effective a given method will perform when used under different circumstances. Validation studies also provide information about how the data derived from these tests should be interpreted. For instance, a validation study assessing a short-term test for potential carcinogens would examine its predictive capability for a random selection of carcinogens and noncarcinogens. The results should indicate how reliable the test is for predicting carcinogenicity, as well as give some indication as to which classes of compounds should or should not be evaluated. There are several ways to quantify the performance of predictive tests, and some of the terms used are defined in Table 3.[28] The three most widely used terms are *sensitivity, specificity,* and *concordance. Sensitivity* refers to the proportion of carcinogens that are positive in a test, while *specificity* is defined as the proportion of noncarcinogens that are negative when tested. *Concordance* is the proportion of chemicals (both carcinogens and noncarcinogens) whose short-term test results agree with *in vivo* data. Since most short-term tests are unable to detect certain classes of carcinogens, the types of compounds in the validation set can dramatically affect the results. An excellent example to illustrate this is the Ames test. Early studies consistently gave predictive success rates of 90% or better. This was to be expected since the Ames test detects mutagens, and these studies used groups of compounds that were predominantly genotoxic carcinogens. Later studies used more balanced selections of genotoxic and nongenotoxic carcinogens and found that sensitivity was as low as 45%.[29,30]

1. Short-Term Test Batteries

The conclusion, that the use of a single short-term test to detect all classes of chemical carcinogens was mechanistically doomed (except when using defined classes of compounds), was established even before it was realized that carcinogens also operate through nongenotoxic mechanisms. Several remedies were proposed, including batteries of tests and tier systems. The practice of using batteries of short-term tests is the most widely used approach and is based on the premise that tests with different endpoints will detect a wider range of compounds. In practice, this has not worked out as well as originally hoped. A major study claimed that the use of a battery of four commonly used tests (Ames test, mouse lymphoma L5178Y cell mutagenesis assay, chromosome aberrations and sister chromatid exchanges) or any combination of these tests, did not substantially improve on the overall performance of the Ames test in predicting rodent carcinogenicity.[29,30] Unfortunately, this study has heightened concerns about the utility of short-term testing for identifying rodent and human carcinogens. Serious reservations about some of the conclusions in the original papers have been delivered.[31] For instance, if the concordance between rat and mouse carcinogenicity results for the first 73 chemicals tested was only 67%. Why did a concordance of ~60% between short-term tests and rodent carcinogenicity discount the predictive utility of short-term tests? These articles[29,31] well document some of the current concepts, drawbacks, and controversies in the field of short-term toxicity testing.

2. Tier Testing

Tier systems involve the use of tests in a hierarchical system where the results at each level provide the basis for more in-depth analysis for the next level.[32] Typically, this might involve a prescreening test such as the Ames test in the first tier to ascertain the genotoxic potential of a compound. A positive response would be followed up in tier two by tests to define the genotoxic potential of the compound in mammalian systems. Tier three testing would assess the risk under defined conditions determined by the results of tier two testing.

3. Decision Point Approach

Another useful way to assess the potential carcinogenicity of a chemical is the *decision point approach* proposed by Williams and Weisburger.[15,33,34] This approach is based on mechanistic knowledge of the chemical obtained in a sequential manner (Table 4). Stage A involves structure-activity studies of the chemical as a preliminary screen for its carcinogenic potential. This is followed by a battery of *in vitro* short-term tests which may also include DNA binding, if possible. The objective of these tests is to identify genotoxic and possibly epigenetic carcinogens at an early stage. After they are performed, decision point 1 is reached, where a decision is made as to the likelihood of the chemical's carcinogenic status. If the results are definitive enough to reach a conclusion that the compound is probably carcinogenic, then no further testing is necessary. Otherwise, the information obtained is used as a guide for further action. For instance, at this point, confirmation of carcinogenicity may be obtained by limited *in vivo* bioassays or, based on the structure and the chemical class of the chemical, tests to detect the potential promoting activity of the chemical may be indicated. As shown in Table 4, several stages of testing are present in which the tests are determined by mechanistic evaluation of results from the previous stages. This approach evaluates potential genotoxic and epigenetic carcinogens in a sequential, scientifically based manner. The information required to make a risk assessment is thus accomplished with a reduced number of procedures and minimal animal use.

Several of the tests used in a decision point approach are outlined in the following sections. These sections give an overview of several of the more well known and useful short-term testing procedures that are a small fraction of the estimated 200 short-term tests already developed.[31] Excellent reviews of many of these tests have been published as a result of the Gene-Tox Program. In this

TABLE 4
Decision Point Approach to Carcinogen Testing

Stage A.   Structure of chemical
Stage B.   Short-term tests *in vitro*
    1.   Mammalian cell DNA repair
    2.   Bacterial mutagenesis
    3.   Mammalian mutagenesis
    4.   Chromosome integrity
    5.   Cell transformation
    Decision point 1: Evaluation of all tests conducted in stages A and B
Stage C.   Tests for promoters
    1.   *In vitro*
    2.   *In vivo*
    Decision point 2: Evaluation of results from stages A through C
Stage D.   Limited *in vivo* bioassays
    1.   Altered foci induction in rodent liver
    2.   Skin neoplasm induction in mice
    3.   Pulmonary neoplasm induction in mice
    4.   Breast cancer induction in female Sprague-Dawley rates
    Decision point 3: Evaluation of results from stages A, B and C and the
    appropriate tests in stage D
Stage E.   Long-term bioassay
    Decision point 4: Final evaluation of all the results and application to health risk
    analysis. This evaluation must include data from stages A, B, and C to provide a
    basis for mechanistic considerations.

From Williams, G. M. and Weisburger, J. H., in *Casarett and Doull's Toxicology: The Basic Science of Poisons*, 4th ed., Amdur, M. O. et al., Eds., Pergamon Press, New York, 1991, 127. With permission.

program, the U.S. Environmental Protection Agency has recently completed two of three phases of its ongoing effort to review the existing literature in chemically induced genetic toxicology. Phase I resulted in the publication of reviews on the evaluation of 36 different bioassays by select panels of experts. Phase II established a database of chemicals analyzed by each group of experts, and phase III is an ongoing effort devoted to the continued review of selected assays and update of the database. The literature resulting from these efforts represents the state-of-the-art in genetic toxicology and also gives an idea as to the diversity of the tests used to detect chemically induced genetic damage. Cimino and Auletta have recently overviewed this program, including an up-to-date bibliography of the publications that it has generated.[35]

II. STRUCTURE-ACTIVITY RELATIONSHIP MODELING

Although obviously not cellular in nature, structure-activity relationship (SAR) modeling can provide the *in vitro* toxicologist with a powerful tool to assess the potential genotoxicity and carcinogenicity of chemicals. SAR modeling attempts to predict the biological activity of a given molecule through analysis of its molecular structure, based on previously analyzed data. In its simplest form, this type of approach entails only an examination of, for example, a small number of PAHs whose potencies range from inactive to highly carcinogenic. Comparison of their structures and accompanying carcinogenic potential might reveal that the "bay region" of the PAH is important to its activity. While the basic principles remain the same, much more powerful SAR modeling becomes possible when computer technology is used to apply statistical analysis between two sets of data. In this comparison the computer program compares the data base containing the biological endpoints of interest to a set of parameters describing the chemical structures. The biological endpoint is then modeled in terms of these parameters. An estimation of the biological activity of an unknown molecule is derived by fitting the parameters describing its chemistry into the model.

Structure-activity relationship models have been used mainly for the development and optimization of pharmaceutical and agricultural products,[36] and only recently have they been used to examine toxicological endpoints such as the ability to induce drug metabolizing enzyme activity.[37-39] Enslein and colleagues have also produced models which are capable of predicting the genotoxic or carcinogenic potential of compounds from several chemical classes.[40-42]

An interesting application of SAR modeling, which has also demonstrated an ability to predict genotoxic and carcinogenic activity, is the computer-automated structure evaluation (CASE) system developed by Rosenkranz and Klopman.[43,44] This artificial intelligence system is unique in that it is completely automated. Instead of the program relying on a series of descriptors selected by the operator, it selects its own fragment descriptors, based on the data base available to the program. For example, the bay region of most PAHs has been shown to be crucial to their carcinogenicity.[45] In most SAR models, if the operator fails to identify an important descriptor to the program, it will remain unrecognized and the predictive value of the model will suffer accordingly. In contrast, when CASE is applied to a data set of PAHs, it will independently identify the importance of the bay region and include it among its descriptors.[43]

SAR models offer valuable alternatives to the use of animals, especially by assisting the researcher to establish priorities in the test evaluation of large numbers of compounds. The value of these models, however, is limited by the data base from which they have been developed. The ideal data base would either consist of randomly selected compounds with proper weighting for various chemical groups, or sufficient numbers of compounds in closely related series in order to permit development of separate models for each group. In practice, most data bases consist predominantly of compounds with safety concerns, i.e., carcinogens which test positively. In addition, many data bases compile information without regard to protocol standardization or the quality and reliability of the test results, thus adding additional uncertainty to formulated conclusions. The availability of data bases such as the U.S. Environmental Protection Agency's Gene-Tox data bank should improve the predictivity of SAR models. This on-line network is available through the National Library of Medicine's TOXNET system and contains genetic toxicology data resulting from expert review of the scientific literature on over 4000 chemicals. Expansion of these types of services is important so that SAR models can be validated with larger numbers of compounds representing a broad spectrum of chemical classes. This effort should reveal the full potential of this approach.

III. METABOLIC ACTIVATION SYSTEMS

A. INTRODUCTION

One of the most common problems confronting *in vitro* toxicologists is the lack of metabolic capacity in the systems with which they are working. This is especially true in the field of chemical carcinogenesis, where it has long been recognized that several environmentally important classes of carcinogens (i.e., the polycyclic aromatic hydrocarbons, *N*-nitrosamines, and aromatic amines) first require metabolic activation in order to exert their ultimate effects.[46-48] These compounds, when tested in most of the short-term tests available today, will exhibit little or no activity unless the assay system has been supplemented with exogenous metabolic activation. Although to date, relatively little attention has been directed towards an understanding of these systems, clearly the choice of the activating system is as important as the test organism or cell in providing the investigator with a definitive outcome.[49]

B. THE S9 ACTIVATION SYSTEM

The first attempts to overcome the obstacle of limited metabolic capacity in short-term test systems (Figure 4), were based on the observation that the metabolism required to activate many carcinogens to their active form was localized primarily with the mixed-function-oxidases in the endoplasmic reticulum of liver cells. Heinrich Malling showed in a bacterial detection system that some promutagens could be activated to their active forms by the addition of a liver homogenate.[11] Subsequently, Bruce Ames and his colleagues[12-14] demonstrated that the predictivity of their bacterial test for carcinogens was substantially increased with the addition of a crude fraction of rat liver homogenate. These findings ensured the place of the *Salmonella/* mammalian-microsome test as the premier method of short-term genotoxicity tests, and also established the hepatic S9 fraction as the activation system of choice in most short-term testing protocols even to the present day. This is despite the fact that the use of cell-free enzyme fractions is well known to be associated with several disadvantages[50] which may in some instances outweigh the advantages (Table 5).

A) SUBCELLULAR ACTIVATION

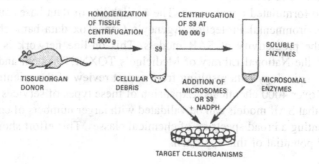

B) INTACT CELLULAR ACTIVATION

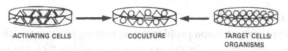

C) HOST-MEDIATED ACTIVATION

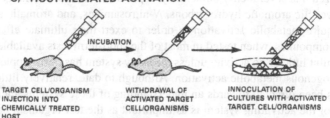

Figure 4.   Different types of metabolic activation for *in vitro* short-term tests.

1. Preparation of S9 Fractions

The activation system used by Ames was derived from the supernatant of the $9000 \times g$ spin (hence, S9) from Aroclor 1254-induced rat liver homogenates. Upon injection in rats, Aroclor 1254 (a complex commercial mixture of poly-chlorinated biphenyls) or a mixture of phenobarbitone and β-naphthaflavone[51] (a lesser health hazard than Aroclor 1254) elevates a broad spectrum of drug-metabolizing enzymes and also induces an S9 fraction with much higher metabolizing activity towards many compounds. Once the induced liver is removed and homogenized, the continuous endoplasmic reticulum fragments to form microvesicles, or microsomes. These microsomes are the primary site of the phase I mixed function oxidase enzymes and are isolated by differential centrifugation of the liver homogenate. The initial step involves centrifugation of the homogenate at $9000 \times g$ to remove nuclei, mitochondria, lysosomes, unbroken cells, and large membrane fragments. The supernatant, or S9, is used in this crude form (the usual practice), or is further enriched for microsomal activity by centrifugation at $100,000 \times g$. The resultant pellet is highly enriched

TABLE 5
Advantages and Disadvantages of Subcellular Activation Systems

Advantages	Disadvantages
Technically easy to prepare in large quantities and easy to use	Toxic to many test systems
Biochemical activity can be altered by varying species, tissue, enzyme inducers, etc.	Requires addition of exogenous cofactors, i.e., NADPH or NADPH-generating system
Moderate cost to prepare	High levels of nucleophiles may bind active molecules and reduce sensitivity
Drug-metabolizing enzymes stable to storage at −70°C	Difficult to identify specific pathways responsible for metabolism of test chemicals
Can be characterized biochemically prior to use	Some phase II (conjugating) drug-metabolizing enzymes not fully represented
Can be filter sterilized (0.45 µM) if aseptic preparation not possible	Enzyme make-up and storage stability not fully characterized
Flexible; can be used with a large variety of systems	Identified as a major contributor to inter-laboratory variation, due to variability in preparation and utilization
Large data base available	Test results very dependent on conditions of use of preparation
Easy to control experimental conditions of exposure	
System favors activation of mutagens/carcinogens and thus may be appropriate for screening studies	
Human studies possible	

for microsomes, while the supernatant fraction, or cytosol, contains many of the enzymes of phase II (conjugative) metabolism.[16] The S9 is incorporated into an S9 mix which includes buffers and cofactors. Detailed descriptions of the preparation of the standard S9 mix are included in several technical reviews.[52,53]

Other activation systems based on enzyme-rich fractions have also been employed. These range from purified enzymes to preparations that attempt to mimic the activation potential of intestinal flora, such as the aptly-named "fecalase" to hydrolyze glycosides,[54] and cell-free extracts of anaerobic bacteria for activating nitroso compounds.[55]

2. Precautions Using the S9 System

The S9 activation system accounts for the majority of variability within the *Salmonella* assay.[56,57] The reasons for this include experimental variations in the enzymes, cofactors, exposure conditions, concentrations of the S9 in the mix, and in the enzyme inducers.[58] Standardization of these parameters appears to eliminate much of this variability.[59] It is important to note, however, that the induced rat hepatic S9 fraction is a complex mixture of detoxifying and activating enzymes whose addition to an assay may not yield results that reflect the species, sex, and organ specificity characteristic of many carcinogens. Thus the "standard" S9 activation system, while experimentally simple and convenient, has limitations which require careful interpretation of experimental results.

Other critics of the utility of the S9 activation system have suggested that S9 preparations do not simulate the *in vivo* activation process and that intact cells may be more representative of *in vivo* conditions and less susceptible to variations within the preparation.[60] In an important series of experiments, Bigger et al.[61,62] examined the binding of the skin carcinogen 7,12-dimethylbenz[a]anthracene (DMBA) to DNA. They suggested that since DNA binding by a chemical carcinogen is considered to be the critical event in the carcinogenic process, the efficacy of various activation systems could be assessed by comparing the DMBA-DNA adducts formed in these systems to the DNA adducts generated in the target skin tissue. In these experiments, DNA was isolated from mouse skin and from various cultured cells exposed to tritiated DMBA, or was recovered from incubation mixtures consisting of calf thymus DNA, tritiated DMBA, cofactors, and rat liver microsomes or S9 fraction. The DMBA-DNA adducts were then compared. It was discovered that the DNA adduct profiles generated in the cell cultures (mouse embryo, human skin, and rat liver) were not qualitatively dependent on the dose of DMBA and were also chromatographically similar to the adducts formed in mouse skin. Conversely, DNA adduct profiles generated using liver homogenates for activation were qualitatively dose-dependent and the chromatograms were markedly dissimilar to that of the target skin tissue. These experiments and others indicate that cellular integrity is important in the metabolism of procarcinogens.[62,63] Although the reasons for this are unclear, it is reasonable to conclude that the organization and spatial orientation of relevant enzymes and cofactors are important for maintaining the integrity of activity, especially since these important cofactors and enzymes may be lost during fractionation.

C. INTACT CELLULAR ACTIVATION SYSTEMS

Intact cellular activation systems are known to be a valuable yet flexible addition to a variety of genotoxic assay systems (Table 6). By combining different cell types and genetic endpoints, the potential genotoxicities of diverse classes of chemicals is determined. Furthermore, the influence of factors such as age, sex, cell type, organ, species, and other variables on genotoxicity can also be elucidated by choosing the appropriate combination of activating and target cells.

A variety of cell types are used for metabolic activation. Although established cell lines are easier to maintain and manipulate in culture, the majority of these lines lack metabolic capacity. Those few cell lines which still maintain some type of drug metabolizing capability exhibit only select enzyme activities. Generally, the activities that are maintained in culture are

TABLE 6
Advantages and Disadvantages of Intact Cellular Activation Systems

Cell type	Advantages	Disadvantages
General	Cellular integrity and cell-cell relationships preserved	More technical difficulty than cell-free preparations
	Endogenous levels of cofactors	More costly than cell-free preparations
	Intermediate between cell-free and *in vivo* activation systems	Limited sample size
		Activation (metabolism) dependent on cell density and preparation quality
Cell Lines	Technically, easiest type of cell culture to handle	Characteristics prone to change with passages in culture
	Generally more hardy to treatment manipulation than primary cultures	Cytochrome P450 (phenobarbitone inducible) enzymes absent
	Generally easiest type of culture to generate large quantities of sample	Other important enzyme systems either absent or present at low levels
	Many different cell lines available which are well characterized, i.e., for cytochrome P448 (3-methylcholanthrene inducible) enzyme activities	
Genetically Altered Cell Lines	Cell lines available with stable expression of inserted enzyme activities, i.e.,rodent and human cytochrome P448 and P450 enzyme activities	Limited access to some lines
		Not representative of complete metabolism
	Good research tool	Not suited to routine screening
		Generally highest technical difficulty
		Many cell types very quickly lose metabolic integrity
		Many cell types not easily amenable to experimental manipulation
Primary Cultures	Excellent research tool, i.e., cell, tissue, and organ specificity	
	Presumably representative of *in situ* cell	
	Human studies possible	

broadly characterized *in vivo* as being inducible by 3-methylcholanthrene, while the enzymes characteristically induced by phenobarbital *in vivo* are lost. The value of using these types of cells as activation systems in predictive genotoxicity testing, therefore, is questionable. Recently, several different forms of the cytochrome P450 genes have been cloned from both rodent and human tissues. Cell lines that exhibit a stable expression of these cytochromes have been used in genotoxicity testing.[64] While this approach is valuable and has demonstrated differences between animal and human forms of cytochrome P450 in their ability to activate promutagens, the constructs are primarily research tools and have limited utility with respect to risk assessment. The genotoxic potential of a chemical to a given target organ is determined by the sum total of the effects of several cytochrome P450s and associated enzymes, and the level of expression depends on the organs and cell types used.

Primary cultures, notably rat hepatocytes, are used extensively as metabolic activation systems in a variety of genetic toxicity assays. Since primary cultures lose metabolic capacity with time in culture and subculture,[65] freshly isolated cells should be used to ensure that their metabolic capability is similar to that found *in situ*. Although it is generally accepted that only qualitative differences exist between animal and human tissues in their ability to metabolize carcinogenic chemicals, the most attractive feature of using primary cultures as activation systems is the opportunity to use human cells. Ironically, any differences between the metabolic capabilities of cultures derived from animals and humans may be masked by the intrinsic variations encountered between different human samples. In addition, the metabolic capacity of "control" human tissue is influenced by the donor's diet, lifestyle, and environment. Nevertheless, very important information can be obtained with respect to the qualitative differences in the metabolism of xenobiotics in animals and humans.

Several models have been developed that incorporate the activation capacity of intact animals prior to expression of chemically induced genetic toxicity in the test organisms *in vitro*. The advantage of these models is that metabolism of the test chemical takes place in the host and avoids the artifacts and inadequacies of *in vitro* activation systems. The intrasanguinous host-mediated assay[66] involves treating mice with the test chemical followed by tail vein injection of two strains of *E. coli,* a wild type and a repair deficient strain. After several hours the bacteria are recovered from different organs and their survival is monitored. The genotoxic effect is measured as the relative survival of the repair deficient vs. the repair proficient strains of bacteria. Similarly, another host-mediated assay involves injection of the test bacterial organisms into the intraperitoneal cavity prior to treatment of the animal with the test substance.[67] After suitable incubation and exposure periods, the bacteria are recovered and the mutagenicity of the test compound is determined. These are two examples of host-mediated assays which can utilize a wide variety of indicator organisms, host species, routes of chemical administration, and genotoxic endpoints within the protocol.[67,68]

Finally, several systems are described in which rats are treated with the test substance for several hours, followed by hepatocyte[69,70] or spermatocyte[71] isolation and culture. During the subsequent 24-h culture period, the cells are exposed to tritiated thymidine, which is incorporated by the nuclei of those cells in chemically induced DNA repair. The repair is then quantitated by autoradiography. The hepatocyte system appears especially promising for the sensitive detection of hepatocarcinogens.

IV. BACTERIAL MUTAGENESIS

A. THE AMES TEST
1. History

Bacterial mutagenesis assays are the most widely used short-term tests for screening of potential mutagens and carcinogens. They combine a high sensitivity for genotoxins with relative technical ease, rapidity, and economy. By far the best known bacterial mutation assay is that developed by Ames and co-workers. Although the *Salmonella* mutagenesis assay was described as early as 1971[72] it was not until Ames introduced the addition of a crude homogenate of rat liver (S9) to the assay system[12-14] that the predictivity of this system for potential carcinogens received serious consideration. In fact, the initial claims of up to 90% prediction rates generated a huge amount of interest that was responsible for the accelerated activity in the development of short-term genotoxicity and carcinogen testing. In contrast to early expectations, however, the predictivity of the *Salmonella*/mammalian-microsome test (commonly referred to as the Ames test) has failed to remain highly predictive when used to evaluate different classes of chemicals.

According to current understanding of chemically induced carcinogenesis, it was premature to expect that one *in vitro* mutagenesis test could have the predictive power for chemical carcinogenicity. It is now recognized that the early success rates were predicated on biased groups of compounds that were already assessed as risks, and that of these, most were genetic carcinogens. In addition, there are several mechanisms of chemical carcinogenesis that mechanistically would not be detected in the Ames test. Nevertheless, the Ames test is the only short-term test considered to be thoroughly validated[28] and remains a very widely used and powerful method for detecting mutagens and potential carcinogens, especially when conducted in a battery of complementary short-term tests. As with all methods, the Ames test is most valuable when the investigator is knowledgeable of its limitations.

2. Experimental Protocol

The Ames test uses several different mutant strains of *Salmonella typhimurium* which are unable to grow under normal conditions unless the cultures are supplemented with histidine. The assay is based on challenging the bacterial cultures with the test compound, followed by incubation for 48 to 72 h on agar medium deficient in histidine. Those bacteria that revert back to histidine independence form colonies which are counted to give an estimate of the mutagenicity of the test compound. The assays are normally conducted with and without S9 activation. This identifies not only those compounds requiring metabolism to express mutagenic activity, but also identifies chemicals which are detoxified by metabolism.

There are a number of *Salmonella* strains that have been specifically designed to detect frameshift or base-pair substitution mutations induced by

TABLE 7

Some Standard Tester Strains Used in *Salmonella*
Mutagenesis Testing

Tester strain	Type of mutation in the histidine gene	LPS[a]	Repair[b]	R Factor[c]
TA 98	Frame shift	*rfa*	*uvrB*	R
TA 100	Base-pair substitution	*rfa*	*uvrB*	R
TA 102	Base-pair substitution	—	—	R
TA 1535	Base-pair substitution	*rfa*	*uvrB*	—
TA 1537	Frame shift	*rfa*	*uvrB*	R

[a]   The *rfa* mutation causes partial loss of the lipopolysaccharide (LPS) coating of
the bacterial cell wall.
[b]   The *uvrB* mutation eliminates DNA excision repair.
[c]   Strains containing the R-factor plasmid (pKM101) have an enhanced error-prone
DNA repair system.

different classes of mutagens.[52] A selection of the most common tester strains
with their mutations is shown in Table 7. All strains contain some type of
mutation in the histidine operon, while additional mutations increase their
sensitivity to mutagens. For instance, the *rfa* mutation causes partial loss of the
lipopolysaccharide surface coating of bacteria and thus increases the perme-
ability to large molecules that do not normally penetrate the cell wall. Another
mutation (*uvrB*), greatly increases the sensitivity of the bacteria towards mutagens by deleting the gene coding for DNA excision repair. Many of the
standard tester strains also contain the R-factor plasmid, pKM101. These R-
factor containing strains are sensitive to a number of mutagens that are inactive
or weakly active in the non-R-factor parent strains. This is apparently due to
the enhancement of an error-prone repair system which is normally present in
these organisms.[73]

The Ames test is a flexible assay which can be adapted to test a large variety
of materials with varying chemical and physical characteristics.[74,75] For routine
chemical screening, however, it is important to adhere to standardized protocols to ensure that the data can be compared to those of other testing laboratories. Several references are available that thoroughly describe standardized
protocols for the Ames test, as well as the preparation of the hepatic S9
fraction.[13,52,53]

Another type of bacterial assay utilizes pairs of DNA repair-deficient and
DNA repair-proficient strains of *E. coli* to test for the genotoxic potential of
chemicals.[76] Although a number of different procedures incorporating different
test organisms are used for this assay, they are all based on the same principles.
Both DNA repair-deficient and proficient strains of the organism are exposed
to several concentrations of the test chemical. After a suitable time interval, the
survival of both strains is assessed. If the survival of the organisms lacking

DNA repair is less than that of the organisms that are DNA repair-proficient, then the chemical is classified as being genotoxic. Although these assays do not test for mutagenicity, they are useful for the detection of potential carcinogens[77] and have been included in a number of screening protocols.

V. MAMMALIAN MUTAGENESIS

A. BASIS OF MAMMALIAN CELL TESTS

Mutagenesis testing in mammalian cells comprises a second tier of genotoxicity testing or screening. In this phase chemicals are tested to confirm that a presumptive mutagen is mutagenic for higher animals. The understanding is that mammalian cells represent a higher degree of organization than bacteria, and that their DNA repair, xenobiotic metabolism, and other related functions also differ markedly. A variety of mammalian cell types and systems have been utilized to screen for chemically induced mutagenic activity.[78] The vast majority of studies, however, have been carried out in three systems which detect mutations at the hypoxanthine-guanine phosphoribosyl transferase (HGPRT) locus using Chinese hamster lung V79 or ovary (CHO) cells, and mutations at the thymidine kinase (TK) locus in the mouse lymphoma T5178Y cell line. The basis of the three tests is similar, although they differ in their experimental design. Chemical treatment of the cells induces a forward mutation at a specific locus, and then the cultures are incubated for a suitable time period to allow phenotypic expression of the induced mutants. These mutants are then allowed to form colonies in medium containing a selective agent which kills wild-type cells and allows only mutants to proliferate. Mutagenic potential is thus quantitated as a function of mutant cell colony formation.

B. FACTORS THAT INFLUENCE PERFORMANCE
OF THE TESTS

Many factors affect the performance of these mutagenesis assays, including the phenotypic expression time, serum, metabolic activation supplementation, and cell density. The phenotypic expression time refers to the interval between the initial mutagenic insult and the maximal expression of the new mutant phenotype. It represents the time required for the mutation to be fixed and for constitutive enzyme and its mRNA to degrade or diminish through cell division. This interval is characteristic of each system and must be carefully determined. Underestimation of the phenotypic expression time results in poor mutant recovery, while longer time periods lengthen the completion of each assay with accompanying increase in cost. Factors such as the doubling time of the cell line, the half-life of the genetic marker, and the selective agent used, all contribute to differences in the phenotypic expression times seen among various systems.[79]

Like most cell lines, V79, CHO, and L5178Y cells lack the metabolic capacity to activate promutagens, and procarcinogens and must therefore be supplemented with an exogenous activation system. This may include hepatic S9 fractions or intact cellular activation.[80] As with all short-term tests, however, the addition of exogenous metabolic activation represents one of the major sources of variation within the testing system. Standardization of metabolic activation procedures within the laboratory can minimize this variation. Serum supplementation of the culture, treatment methods, and selection medium[81-83] can also contribute to variations in test results. Sera can vary markedly from lot to lot, and affect not only the growth characteristics of the cells, but also the activity of the test agent in the assay. Different lots of serum should be screened to ensure that the phenotypic growth characteristics of the cultures are maintained at acceptable levels. The assay should also be performed with "standard" mutagens as an internal experimental control (see Chapter 1, Section V, for discussion of the use of serum supplemented media).

One of the more interesting examples of factors that can profoundly affect the performance of these assays is cell density. It has been shown that cell-cell contact can reduce the number of mutants recovered due to intercellular communication, especially with HGPRT mutant selection.[84-87] This "metabolic cooperation" occurs during 6-thioguanine selection and involves the transfer of toxic metabolites (6-thioguanine monophosphate) through membrane gap junctions of the dying wild-type cells to the healthy mutants. Thus, the activation of 6-thioguanine is lethal for the mutant cells due to its metabolism in the parental cells and subsequent transfer of the toxic enzyme metabolite. The parent compound, however, is not toxic to the mutants. Maintaining the cultures at densities where cell-cell contact is minimized during mutant selection prevents metabolic cooperation.

Although metabolic cooperation can confound mutagenicity studies, interestingly, it is used as the basis to develop bioassays for detecting potential epigenetic carcinogens, i.e., cocarcinogens and promoters. Loss of cell-cell communication may cause major changes in cellular differentiation, which has been implicated in the expression of cancer. Consequently, quantification of the effects of agents which cause the loss of metabolic cooperation *in vitro* may provide a measure of potential carcinogenicity. Experiments have clearly shown that several tumor promoters and epigenetic carcinogens,[89,90] including TPA,[88] inhibit metabolic cooperation. It appears that the inhibition of metabolic cooperation *in vitro* is associated with tumor promotion, cocarcinogenesis, and perhaps teratogenesis *in vivo*.[89-91]

C. THE CELL LINES
1. CHO and V79 Cells

The CHO and V79 cell lines are used with a variety of genetic markers, including resistance to ouabain, 8-azaadenine, 2,6-deaminopurine, emetine, alpha-amanitin, cycloheximide and methotrexate, to detect chemical mutagens.[92,93] The best characterized mutagenesis system in either cell line, however, involves mutation at the HGPRT locus. This locus encodes for the

HGPRT enzyme, which is normally responsible for the conversion of the purines hypoxanthine and guanine into nucleotides using a purine salvage pathway. This gene is present on the X chromosome of humans, and HGPRT deficiencies lead to disorders of uric acid metabolism and central nervous system function (Lesch-Nyhan syndrome). The assay relies on the metabolism of purine analogs, such as 6-thioguanine and 8-azaguanine, by HGPRT to toxic ribophosphorylated derivatives, which cause cell death when incorporated into DNA. Mutations at the HGPRT locus inactivate the enzyme and inhibit the catalytic formation of toxic metabolites from 6-thioguanine. The mutant cells are able to survive using *de novo* purine biosynthesis and form colonies at concentrations of 6-thioguanine normally cytotoxic to wild-type cells.

2. Mouse Lymphoma L5178Y Cells

Mouse lymphoma L5178Y cells are also used in mutagenesis assays that utilize a variety of genetic markers. The best-characterized system is based on the quantitation of forward mutations occurring at the heterozygous TK locus.[94] The TK enzyme is responsible for the incorporation of exogenous thymidine into the cell in the form of thymidine monophosphate. When the cell uses trifluorothymidine (TFT) as the substrate instead of thymidine, the resultant phosphorylated TFT acts as an irreversible inhibitor of thymidylate synthetase, resulting in cell death. Thus, in the L5178Y mutagenesis system, when lymphoma cells are exposed to a putative mutagen, forward mutations occur at the TK locus and inactivate the TK enzyme. These mutants are then selected for by the inclusion of TFT in the culture medium. Wild-type cells are killed, while TK mutants survive by synthesizing purines *de novo*.

An interesting aspect of the TK mutagenesis assay is the possibility that it may also simultaneously provide information on the clastogenic potential of a test compound, i.e., the induction of chromosome breakage. It has long been recognized that chemical treatment of L5178Y cells gives rise to the growth of colonies with a range of sizes on the TFT selection plates. Although the significance of labeling the small colonies as *bona fide* mutants was initially questioned, more recent work has confirmed that these small colonies are indeed TK mutants.[95] Furthermore, a strong correlation exists between the small colony TK mutants and chromosome damage.[96,97] Whether all clastogens, or only specific classes of clastogens, induce these small colony mutants is still being investigated. Nevertheless, this assay has the potential to detect both the mutagenic and clastogenic effects of chemicals.

All three assays are relatively technically easy to perform, are sensitive to most genotoxic chemicals, but lack metabolic capacity. The Chinese hamster cell systems are also used extensively to examine chromosome aberrations and sister chromatid exchanges, thus allowing for the comparison of tests having different endpoints. Conversely, the mouse lymphoma cell system has the potential for the simultaneous assessment of both the mutagenic and clastogenic potential of test compounds. It does produce false positive results, however, which detracts from its usefulness. The theory and methodology of the V79/HGPRT,[93,98] CHO/HGPRT,[92,99–102] and L5178Y/TK[94,96,103–105] systems are described in several references. In addition, Nestmann et al.[106] have compared the three assays, detailing the protocols and identifying the various advantages and disadvantages of each assay.

VI. *IN VITRO* CYTOGENETIC TESTING

A. INTRODUCTION

In contrast to the mutagenesis assays, cytogenetic assays are used to determine chemically induced changes in the structure or number of chromosomes, as seen through the light microscope. Many techniques are available for cytogenetic testing both *in vivo* and *in vitro*, and they are divided into three different types:

1.   Tests to detect chromosomal aberrations
2.   Tests to detect exchanges of chromosomal material between sister chromatids
3.   Tests to detect chromosome fragments, or "micronuclei" resulting from chromosome damage

In general, these techniques were originally developed for *in vivo* testing, but with the recent and rapid advances in tissue culture technology, *in vitro* techniques have been developed which are more convenient and economical to screen large numbers of compounds. A variety of *in vivo/in vitro* procedures are also available where the chemical exposure takes place *in vivo*, followed by the culturing of peripheral lymphocytes. This approach takes on added significance since it is one of the few methods available to monitor human exposure to genotoxic agents.

B. CHROMOSOMAL ABERRATIONS

Chromosomal aberrations are implicated in the induction of congenital abnormalities and cancer. Most known carcinogens have also been identified as "clastogens", or agents that induce chromosome breakage.[107] *In vivo* methods generally involve treatment of the animal with the test agent, followed by microscopic examination of different cell types such as bone marrow, spermatogonia, spermatocytes, oocytes, and early embryos.[108] *In vitro* methods are more sensitive than *in vivo* methods but they usually require supplementation with exogenous activation. For short-term cytogenetic studies, however, cell cultures are the systems of choice,[108] where CHO and V79 cell lines and human lymphocytes are more commonly used.

The basic protocol for *in vitro* detection of chromosome aberrations is similar for most cell types. Cultured cells that are not normally proliferative, such as lymphocytes, are stimulated to enter the cell cycle by the addition of a mitogen, such as phytohaemagglutinin (PHA). Since the induction of cell aberrations is contingent on active cell division, the mitotic index or other measure of cytotoxicity is determined at a preliminary range of test chemical doses to ensure that the final dosing schedule does not include overly cytotoxic doses. The cells are then treated for sufficient time to ensure that the cells are exposed to the test chemical during all stages of the cell cycle. Alternatively, the cells are harvested at different time intervals to examine the effects of test chemical at different stages. If S9 activation is needed, the chemical exposures are reduced to 2 to 3 h, since S9 preparations are cytotoxic. The cultures are then washed, and colchicine is added 2 to 3 h prior to harvesting the cultures, to arrest the cells in metaphase for analysis of chromosome aberrations.

Successful "metaphase spreads" suitable for scoring are achieved with experience and patience. The procedure entails careful swelling of the cells by the addition of hypotonic potassium chloride solution. The swollen cells are fixed and then dropped onto precleaned, cold, wet slides, where the impact explodes the turgid cells onto the slide and spreads the metaphase chromosomes enough so that microscopic examination of each chromosome is possible after staining. Genotoxic chemicals cause different types of structural aberrations and several examples are shown in Figure 5.[109] Both numbers and types of aberrations are scored and the assessment of these tests is a painstaking process requiring considerable patience and expertise.

C. SISTER CHROMATID EXCHANGE

Sister chromatid exchange (SCE) involves the interchange of replicating DNA between chromatids at apparently homologous loci. Unlike chromosome aberrations, this exchange involves no overall change in morphology and can be detected only by differential labeling of the sister chromatids. Original methods used tritiated thymidine, where chromosomes were allowed to replicate once in the presence of the label and then again in its absence. Upon autoradiograph exposure, only one of the sister chromatids would show the radioactive label, due to the semiconservative nature of DNA replication. Taylor named the occasional symmetrical switches in this label between daughter chromatids "sister chromatid exchange".[110] Further work showed that their frequency could be increased in a dose-dependent manner by exposure to a host of physical and chemical mutagens.

Newer methods use 5-bromodeoxyuridine (BrdUrd) label combined with different stains. The resolution of these methods is higher and the detection of SCE with these techniques is one of the quickest, easiest, and most sensitive tests for chemically induced genetic damage.[111] The protocols are similar to those used to detect chromosome aberrations, except that the BrdUrd label is added to the culture medium so that the cultures undergo at least two replications in its presence prior to harvesting. The slides are stained after harvesting and spreading of the metaphase chromosomes. A popular staining technique to distinguish sister chromatids is the fluorescent plus Giemsa (FPG) method

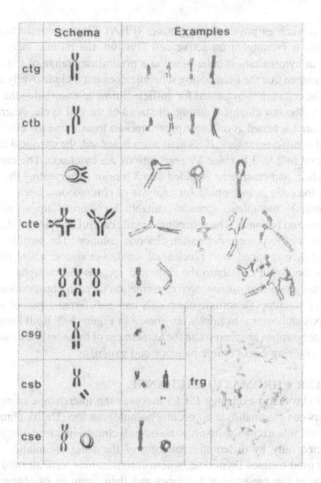

Figure 5. A diagram and examples of different types of structural chromosomal aberrations including chromatid gaps (ctg), breaks (ctb), and exchanges (cte); and chromosome gaps (csg), breaks (csb), and exchanges (cse). (From Ishidate, M., Jr., *Data Book of Chromosomal Aberration Test In Vitro*, rev. ed., Elsevier, Amsterdam, 1988. With permission.)

which produces differentially stained, "harlequin" chromosomes.[111] It has the advantage of producing permanent cytological preparations that are viewed with a light microscope. Figure 6 shows the induction of SCE in CHO cells by adriamycin.[112]

Although the biological significance of SCE and the molecular mechanisms that produce them are not understood, a strong correlation has been shown between the induction of SCE and carcinogenicity.[107] In addition, SCE testing is considerably less labor intensive than tests for chromosome aberrations. Together this makes SCE detection a useful short-term test for carcinogens and a sensitive indicator of genotoxic exposure when detected in human peripheral lymphocytes.[113]

D. DETECTION OF MICRONUCLEI

Other short-term cytogenetic tests that combine sensitivity with simplicity are those that measure the production of micronuclei. Micronuclei arise from chromosomal fragments or chromosomes that are not passed to the daughter nuclei during cell division. The sensitivity of the micronucleus test is comparable to those measuring chromosome aberrations. The micronucleus test, however, detects only chromosome breakage, while the latter test identifies events such as chromosome breakage, exchanges and translocations.

The measurement of micronuclei is faster and easier than scoring chromosome aberrations,[114] and micronucleus assays are used with a variety of organisms and cells, including plants, animal and human cells in culture, and whole animals.[115] Systems utilizing bone marrow from whole animals[116] are widely used (theoretically, any dividing cell population is a candidate). The *in vitro* protocols are similar to those used for chromosome aberrations. The major difference is that the dividing cell population is treated with cytochalasin B to prevent cytokinesis.[117] Nuclear division is not blocked and cytochalasin B-treated cells that have two nuclei are scored for the presence of micronuclei that are not incorporated into either daughter nucleus.

As with many other short-term testing systems, the major disadvantage of *in vitro* cytogenetic systems is their lack of metabolic integrity. Microsomal enzyme preparations and feeder layers are used to partially circumvent this deficiency. The tests for chromosome aberrations are the only tests that provide definitive identification of different kinds of chemically induced chromosomal damage, but they are time consuming and require considerable expertise. Alternative tests, such as the micronucleus test or the extremely sensitive SCE test, do not provide as much information, but are much easier to carry out, are faster, and are more economical. Useful references describing the theory and protocols for carrying out chromosome aberration tests,[108,109,118] tests for SCE,[118-122] and micronucleus tests[114-116] can be found in the reference section.

VII. UNSCHEDULED DNA SYNTHESIS

A. INTRODUCTION

Unscheduled DNA synthesis (UDS) quantitation is an indirect measurement of DNA damage. These assays are based on the understanding that DNA repair is a sequential enzymatic process initiated by the cell's recognition of the presence of DNA chemical adducts. The original primary structure of the DNA is restored through excision of the adducts, DNA strand polymerization using the opposing strand as a template, and subsequent ligation.[123] By incorporating a radioactive nucleotide base (usually thymidine) into the base pool during the repair process, the repaired lesion is radiolabeled and this "unscheduled DNA synthesis" can then be quantitated by autoradiography or liquid scintillation (LC) counting. Since a large proportion of chemical mutagens and carcinogens

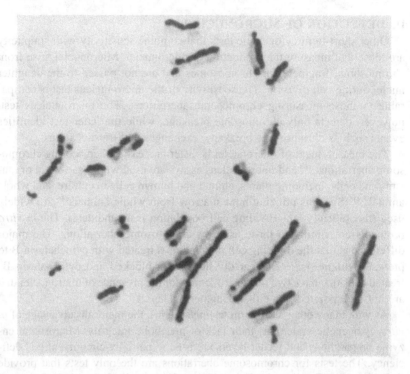

Figure 6a. Chinese hamster ovary cells cultured for two cell cycles in the presence of 10 μM 5-bromodeoxyuridine and stained with the fluorescent plus Giemsa technique. Figure 6a shows a control cell with 11 SCEs. Figure 6b shows a cell treated with 3×10^{-7} M adriamycin for 24 h that has 72 SCEs. (From Perry, P.E., in *Chemical Mutagens*, Vol. 6, de Serres, F.J. and Hollaender, A., Eds., Plenum, New York, 1980, 1. With permission.)

have been shown to damage DNA, UDS provides a sensitive, economic, and relatively easy method with which to detect potential chemical genotoxicity and carcinogenicity.

B. RADIOACTIVE QUANTITATIVE METHODS

The simplest and quickest way to measure UDS entails LC quantitation of radioactivity associated with acid precipitable material.[124] The sensitivity of this method, however, is low due to high background counts associated with the acid precipitable fraction. Other procedures have been developed that attempt to reduce the background by isolating either nuclei[125,126] or DNA[127] prior to LC counting. The radioactivity is then quantitated and UDS is expressed as disintegrations per minute of incorporated label per unit (μg) of DNA. Although the assay is capable of quickly screening large numbers of compounds, counts that result from replicative DNA synthesis are not distinguished from those that result from repair synthesis. Furthermore, S-phase cells incorporate more radioactivity which may mask the UDS response. This is a serious problem which leads to false positive results. With

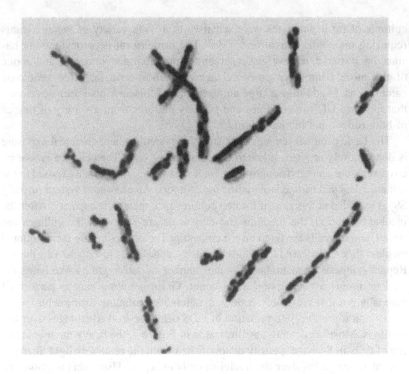

Figure 6b.

proliferating cell cultures it is necessary to block replicative DNA synthesis with chemical inhibitors or arginine deficiency. Some difficulties, however, arise with the use of hydroxyurea and aphidicolin as inhibitors of replicative DNA synthesis. The former chemical stimulates DNA repair, while the latter causes a slight, but significant, inhibition of DNA repair in human epidermal cells.[128]

C. AUTORADIOGRAPHIC METHODS

Autoradiography is the most common technique for measuring UDS. Although this method is somewhat more technically demanding and more time consuming than LC methods, it is generally accepted as having better reliability and sensitivity. Rasmussen and Painter first described studies using autoradiographic techniques to measure DNA damage in mammalian cells. They showed that ultraviolet light induced the uptake of labeled thymidine into the DNA of cells not in replicative DNA synthesis.[129,130] This approach was quickly applied to chemically induced damage, and by the early 1970s UDS was in place as a screen for chemical genotoxicants.[131] These early studies, however, used cell types with little metabolic capacity and were not sensitive to indirect genotoxic agents. Although metabolic capacity can be added to these systems in the form of hepatic S9 fractions, knowledge in this area advanced dramatically with the demonstration that freshly isolated primary cultures of rat hepatocytes were sensitive to a wide variety of genotoxicants requiring metabolic activation.[125,132,133] The primary rat hepatocyte assay has since been used extensively as a screen for potential genotoxicants and is one of the more commonly used and accepted short-term tests for genotoxic carcinogens. In addition, a large number of protocols have also been developed that measure UDS in cell lines and primary cultures from a variety of tissues of both rodent and human origin.[134,135]

The basic protocol remains consistent for systems where chemical exposure is done entirely *in vitro*, although the details may vary. The cells are grown on coverslips, are rendered nonproliferative if necessary, and are then exposed to the test chemical and tritiated thymidine for 24 hours. An activation system (usually S9) is included at this step, if the cell cultures lack metabolic capacity. After the incubation period, the medium and chemicals are removed, the cultures are rinsed, and the cells are fixed on the coverslip. The coverslips are then mounted on glass slides, processed for autoradiography, and stained to visualize the nuclei. Repair synthesis is quantitated as the number of silver grains overlying the nucleus minus a background grain count. Grain counting can be performed manually but it is preferably done with automated counting instruments.

Autoradiographic determination of UDS offers several advantages over LC methods. Most importantly, as illustrated in Figure 7, the heavy nuclear labeling of cells in S-phase is easily distinguished from the relatively light distribution of silver grains over the nuclei of cells in repair. Using this method, the UDS response is quantitated as both grain counts and the percentage of the cell population in UDS. Cytotoxic concentrations of test chemicals are also evaluated by visual inspection of the cells prior to autoradiography. Cytotoxicity has been found to be a confounding factor in many short-term assays. UDS is dose dependent however, and occurs with noncytotoxic concentrations of the chemical. In fact, as the dose becomes cytotoxic, UDS is inhibited.[136]

A variation of the rat hepatocyte/UDS system takes advantage of the metabolism of the intact animal. In this *in vivo/in vitro* model the rat is treated with the test chemical from 1 to 16 h prior to sacrifice. Hepatocytes are cultured, and labeling is initiated as soon as the cells have anchored to the culture surface. After a short labeling period (4 h) the cultures are rinsed and refed with medium containing nonradioactive thymidine. After a further 16-h incubation, the cultures are processed for autoradiography. This procedure yields results which reflect the absorption, distribution, biotransformation, and elimination of the test compound *in vivo*. It also contributes to understanding the influence of different factors such as species, strain, sex, chronic exposure, and route of exposure on genotoxicity. In addition, any effects of the test chemical on hepatic cell proliferation can be quantitated. Although the exact role of cell proliferation in hepatic carcinogenesis is not known, it appears that chemically induced cell proliferation may serve as a form of tumor promotion for livers that have already been initiated by genotoxic agents. Measurement of induced cell proliferation may provide a useful tool for assessing the potential hepatotoxicity and carcinogenicity of a compound.[137]

UDS is generally accepted as a reliable and sensitive measure of genotoxic activity,[138,139] and several studies examining the *in vitro* hepatocyte/UDS system have supported this consensus.[140-142] The measurement of UDS in mammalian cells provides a relatively easy and inexpensive method for the detection of potentially hazardous chemicals that have mutagenic or carcinogenic properties. Since UDS measures DNA damage over the entire genome and is sensitive only to genotoxic compounds, its inclusion in a battery of tests with broader specificity allows for the identification of compounds that may be active through epigenetic mechanisms. Several references describe the theory and methodology for both the *in vitro*[143,144] and the *in vivo/in vitro*[70,144] rat hepatocyte UDS systems. In addition, Swierenga and colleagues[145] describe the hepatocyte/UDS autoradiographic method as it is currently used. This report details the UDS methodology and the authors insert useful comments with each step providing a guide for investigators who intend to use autoradiographic detection of UDS *in vitro*.

VIII.   CELL TRANSFORMATION

A. INTRODUCTION

In vitro models of cell transformation are perhaps the most mechanistically appropriate short-term tests for the complex sequence of events leading to chemically induced cancers *in vivo*. Cells that become malignantly transformed in these tests have certain phenotypic alterations that are related to cancer. The tests use a variety of endpoints, based on these alterations, to detect the transforming ability and carcinogenic potential of test compounds. These endpoints include loss of anchorage dependence, alterations in morphology or behavioral characteristics, i.e., loss of contact inhibition, altered growth in agar, viral dependence and the ability to induce cancer when transplanted into syngeneic host animals.[146,147]

Chemically induced malignant transformation *in vitro* was first reported in 1943 when Earle and Nettleship showed that mouse embryo cells became transformed when exposed to PAHs.[148] They also found, however, that untreated control cells became transformed. These results effectively delayed the understanding underlying transformation of cells in culture in response to chemical carcinogens. In 1963, Berwald and Sachs rediscovered the phenomenon, using primary cultures of Syrian hamster embryo (SHE) fibroblasts[149] and subsequently showed that spontaneous transformation in this species occurred much less frequently than with mouse embryo cells.[150] A quantitative colony assay was described in which normal cells formed colonies with monolayer, epithelial-like characteristics, and transformed colonies exhibited a multilayer, fibroblastic morphology, with disordered and distorted growth patterns.[151] Subsequent studies in other systems with a variety of carcinogens revealed that cell transformation was a legitimate endpoint for the assessment of potential chemical carcinogens, and laid the groundwork for

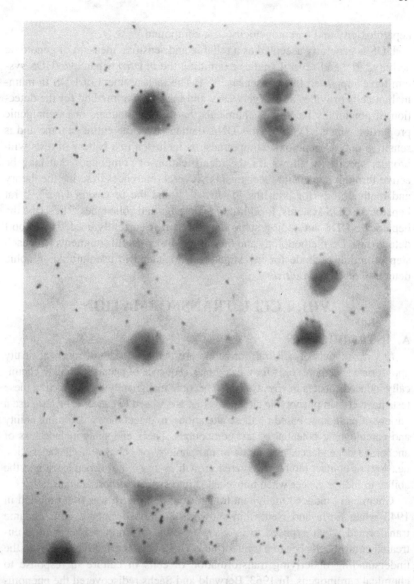

Figure 7a. *In vivo/in vitro* unscheduled DNA synthesis in mouse hepatocytes. Mice were treated orally with either saline (a) or 20 mg/kg dimethylnitrosamine (b). Two hours later the livers were removed and the hepatocytes were cultured and incubated in the presence of tritiated thymidine for 4 h. Note the heavy labeling of the two S-phase cells in replicative DNA synthesis (b). (Courtesy of Drs. B. Butterworth and J. Larsen).

future developments in the field.

Despite this promising beginning, in some respects, cell transformation assays remain problematic among short-term screening tests. Results are notoriously difficult to reproduce among different testing laboratories, and a large degree of

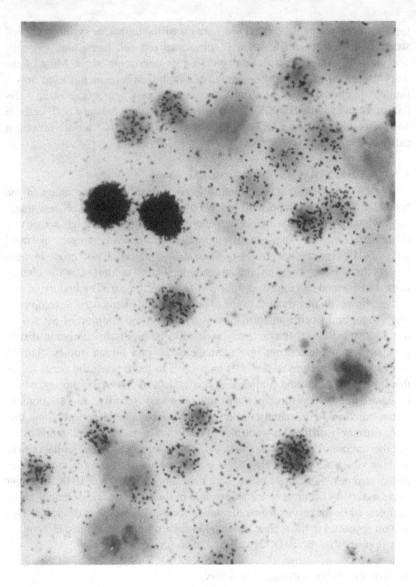

Figure 7b.

variability is not uncommon within the same laboratory. Careful assessment of each cell culture preparation prior to testing is necessary, as well as the strict standardization of protocols. The variations in the data are due, in part, to the complexities of the transformation process and to the subjectivity of the scoring.

B. EXPERIMENTAL MODELS
Many models have been developed which utilize a number of different cell

types and end points. In general, they are classified into three systems: (1) cell strains with a limited life span; (2) immortalized cell lines; and, (3) cells concomitantly infected with or expressing an oncogenic virus. Most of the work using cell transformation assays as carcinogen screens has used only a limited number of systems. These include the soft agar assay using baby hamster kidney cells (BHK 21),[151] colony assays using primary SHE cells,[152] and focus assays using BALB/c 3T3[153] or C3H 10T1/2[154,155] mouse fibroblast cells.

1. The BHK Soft Agar Transformation Assay

The BHK soft agar transformation assay was unique since many of the promotional processes appeared to have already taken place. The near-transformed state of the cells before treatment thus conferred a high sensitivity towards carcinogens. The assay itself appeared deceptively simple; "normal" untreated cells were anchorage-dependent and thus did not grow in agar (except for a background of anchorage-independent colonies), while chemically transformed cells grew and formed colonies in agar. The history of this system, however, illustrates the type of problems that plagued the development and validation of cell transformation assays. With the addition of S9 activation,[151,156] the BHK soft agar assay seemed to be highly predictive in distinguishing carcinogens from noncarcinogens. In fact initial studies claimed success rates as high as 91%,[157,158] comparable to those using the Ames test at that time. This prompted studies with this system in other laboratories, where it seemed that its performance was not only very sensitive to variations in serum batch and concentration, but also to environmental perturbations that were extremely difficult to control.[159] The difficulty in transporting this assay to other laboratories was demonstrated during an international collaborative study in which 42 selected chemicals were sent to participating laboratories as coded samples.[160] The results of the parent laboratory of the BHK soft agar assay were very accurate in the assessment of the carcinogenic potential of the samples, while the results from other laboratories were less than desirable. This system appeared to have great potential in spite of these problems, although it now appears that the BHK 21 cell line is no longer available.[90]

2. The Focus Transformation Assay

The focus transformation assays incorporate the aneuploid C3H/10T1/2 and BALB/c 3T3 cell lines within protocols that are similar to each other (Figure 8). The cell lines have a normal fusiform morphology, display density-dependent inhibition of growth, and form foci on top of the monolayer in culture and are tumorigenic *in vivo* when transformed. With minor variations, the actual assays consist of seeding relatively small numbers of cells from frozen stocks on day one. The next day the cultures are exposed to the test chemical for 24 to 72 h (if exogenous metabolic activation is added at this stage, only a 2- to 4-h treatment period is used). The cultures are then incubated for an additional 4

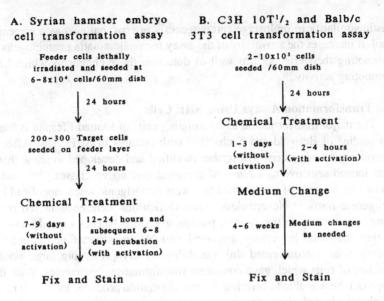

Figure 8. Schematic representations of protocols for the SHE (A) and C3H 10T1/2 and BALB/c 3T3 (B) cell transformation assays.

to 6 weeks to express the transformed phenotype. The cultures are fixed and stained to determine the presence of foci of transformed cells. The scoring of these foci is susceptible to some degree of subjectivity and is undoubtedly the source of some variation among results from different laboratories. Three types of foci are recognized in these assays, designated as types I, II, and III.[161] Type I is characterized by tightly packed but monolayered cells which have not been found to be tumorigenic after inoculation into host animals, and are therefore not scored as malignantly transformed. Type II foci, not normally present in significant numbers in BALB/c 3T3 cultures, exhibit a dense multilayer network of cells with well-ordered and defined edges. The overlapping of cells is evident in these foci, but not pronounced. Type III foci are densely staining groups of multilayered cells having a pronounced overlapping pattern of growth. They are also distinct from type II foci in that the edges are irregular and are markedly more tumorigenic when injected into syngeneic host animals.[154]

An interesting development was the demonstration that the procedures could be modified to mimic the two-stage initiation-promotion model of chemical carcinogenesis that is well-documented in mouse skin *in vivo*.[162,163] Operationally, the *in vitro* system is very similar to its *in vivo* counterpart. The cultures are "initiated" with single, subthreshold doses of a carcinogen, such as 3-methylcholanthrene or DMBA, and then are treated repeatedly with a promoter such as TPA. Either treatment regimen by itself induces the formation of few, if any transformed foci. The complete protocol, however, produces large numbers of foci. This approach is extremely valuable not only from its mechanistic similarity to multistage carcinogenesis *in vivo*, but also as a screening tool. It increases the sensitivity of the assay for weak initiating carcinogens by promoting their effects,[164] as well as detecting those chemicals which have promoting activity.[90]

3. Transformation Assays Using SHE Cells

The major transformation assay utilizing cells of limited lifespan is based on studies of Berwald and Sachs[149,150] with cultures of SHE cells. After its introduction, the system was further modified and developed to show that it was indeed sensitive to a range of chemical carcinogen classes,[165,166] and to show that the cells of transformed foci were tumorigenic when inoculated into syngeneic hosts.[167] Nevertheless, many difficulties were encountered in attempts to reproduce the system, particularly since batches of cells handled within the same laboratory displayed variable transformation frequencies. Pienta et al. circumvented this variability by cryopreserving large pooled batches of cells which gave consistent transformation frequencies. With this approach he was able to correlate the transformation produced by a wide range of chemicals with their carcinogenic activities *in vivo*.[152]

Transformation assays using SHE cells have employed both focus assays[168] and clonal assays. In the clonal assay (Figure 8) a small number (300 to 500) of target cells are seeded in petri dishes previously inoculated with 60,000 to 80,000 lethally irradiated feeder cells. Both target and feeder cells are derived from either freshly prepared or cryopreserved SHE preparations. The next day, the cultures are exposed to the test chemical and incubated for an additional 7 to 9 days. The medium is then removed and the cultures are washed, fixed, stained, and examined for morphologic transformation. The transformed colonies of this assay are discrete, separate entities as compared to those of the cell line assays outlined above. The major advantage of this method is the relatively short time period required to complete the tests.

The focus assay has been used to evaluate only a limited number of test compounds. This procedure incorporates secondary cultures of target SHE cells seeded at a higher density (50,000 cells/50 mm dish) than in the clonal assay without the use of a feeder layer. The cells are allowed to incubate for 3 days prior to a 3-day chemical treatment. After an additional 20 to 25 days, the cultures are processed for evaluation of transformed foci.

Although early passage SHE cells retain significant levels of drug metabolizing enzyme activity even after cryopreservation, they lack the enzyme systems necessary to activate several classes of carcinogens. It is common practice not to include exogenous metabolic activation in cell transformation systems. It is advisable to do so, however, in order to decrease false-negative responses.[50,169]

4. Viral-Chemical Transformation Methods

Several systems are predicated on viral-chemical interactions to measure transformation. Although they are used extensively in mechanistic studies, they have also found utility as short-term screening tests. Examples include the SHE/SA7 transformation assay,[170,171] which is based on the observation that DNA-damaging agents will increase the transformation frequencies of simian adenovirus (SA7)-infected SHE cells, and the FRE/RLV transformation assay,[172] where Fisher rat embryo (FRE) cells, which are normally insensitive to chemical transformation, are rendered sensitive by infection with Rauscher leukemia virus (RLV).

In vitro cell transformation assays are powerful tools which are used to study the mechanisms of chemical carcinogenesis, or as short-term tests to screen for potential carcinogens. A particularly attractive characteristic of these procedures is their sensitivity to both genotoxic and epigenetic carcinogens. Their overall performance, therefore, as predictors of potential carcinogens, is superior to many commonly used short-term tests.[90] It is important to note however, that the procedures are technically difficult and their performance is highly variable, even when the most rigid experimental conditions are closely followed.[147,159,173,174] Several references describing the theory and procedures of the C3H/10T1/2,[161,175,176,178,179] BALB/c 3T3,[161,175–177,179] and SHE[161,175] clonal assays are useful for further investigation and understanding of the concepts involving cell transformation systems.

IX. SUMMARY

Short-term tests offer the researcher a convenient and economical means to assess the genotoxicity and/or potential carcinogenicity of chemicals. They are widely employed in mechanistic studies, and have increasingly become routine tools to screen large numbers of compounds for regulatory and commercial purposes. As with all experimental protocols, the application of short-term testing procedures is most successful when applied in a manner consistent with their scientific basis. Similarly, interpretation of the data obtained from these tests should be made within the limitations of the assay. As our understanding of the endpoints increases, they will become more powerful tools for the detection of chemically induced genotoxicity and carcinogenicity.

REFERENCES

1. **Pott, P.**, Chirurgical observations relative to cancer of the scrotum, London, 1774, reprinted in *Nat. Cancer Inst. Monogr.*, 10, 7, 1963.
2. **Yamagiwa, K. and Ichikawa, K.**, Experimentelle Studie über die Pathogenese der Epithelialgeschwulste, *Mitt. Med. Fak.*, Tokyo, 15, 295, 1915.
3. **Yamagiwa, K. and Ichikawa, K.**, Experimental study of the pathogenesis of carcinoma, *J. Cancer Res.*, 3, 1, 1918.
4. **Cook, J.W., Hieger, I., Kennaway, E.L., and Mayneord, W.V.**, The production of cancer by pure hydrocarbons. I. *Proc. R. Soc. London B*, 111, 455, 1932.
5. **Boyland, E.**, Chemical carcinogenesis and experimental chemotherapy of cancer, *Yale J. Biol. Med.*, 20, 321, 1947.

6. **Miller, J.A. and Miller, E.C.,** The metabolic activation of carcinogenic aromatic amines and amides, *Progr. Exp. Tumor Res.*, 11, 273, 1969.

7. **Boveri, T.,** Zur Frage der Enstehung maligner Tumoren, *Jena*, 1914.

8. **Müller, H.J.,** Artificial transmutation of the gene, *Science*, 66, 84, 1927.

9. **Auerbach, C. and Robson, J.M.,** Chemical production of mutation, *Nature*, 157, 302, 1946.

10. **Brookes, P. and Lawley, P.D.,** Evidence for the binding of polynuclear aromatic hydrocarbons to the nucleic acids of mouse skin: relation between carcinogenic power of hydrocarbons and their binding to deoxyribonucleic acid, *Nature*, 202, 781, 1964.

11. **Malling, H.V.,** Dimethylnitrosamine: formation of mutagenic compounds by interaction with mouse liver microsomes, *Mutat. Res.*, 13, 425, 1971.

12. **Ames, B.N., Durston, W.E., Yamasaki, E., and Lee, F.D.,** Carcinogens are mutagens: a simple test system combining liver homogenates for activation and bacteria for detection, *Proc. Natl. Acad. Sci. U.S.A.*, 70, 2281, 1973.

13. **Ames, B.N., McCann, J., and Yamasaki, E.,** Methods for detecting carcinogens and mutagens with the *Salmonella*/mammalian-microsome mutagenicity test, *Mutat. Res.*, 31, 347, 1975.

14. **McCann, J., Choi, E., Yamasaki, E., and Ames, B.N.,** Detection of carcinogens as mutagens in the Salmonella/microsome test: assay of 300 chemicals, *Proc. Natl. Acad. Sci. U.S.A.*, 72, 5135, 1975.

15. **Williams, G.M. and Weisburger, J.H.,** Chemical carcinogenesis, in *Casarett and Doull's Toxicology: The Basic Science of Poisons*, 4th ed., Amdur, M.O., Doull, J., and Klaassen, C.D., Eds., Pergamon Press, New York, 1991, 127.

16. **Sipes, I.G. and Gandolfi, A.J.,** Biotransformation of toxicants, in *Casarett and Doull's Toxicology: The Basic Science of Poisons*, 4th ed., Amdur, M.O., Doull, J. and Klaassen, C.D., Eds., Pergamon Press, New York, 1991, 88.

17. **Phillips, D.H., Grover, P.L., and Sims, P.,** A quantitative determination of the covalent binding of a series of polycyclic hydrocarbons to DNA in mouse skin, *Int. J. Cancer*, 23, 201, 1979.

18. **Randerath, K., Randerath, E., Agrawal, H.P., Gupta, R.C., Schurdak, M.E.. and Reddy, M.V.,** Postlabeling methods for carcinogen-DNA adduct analysis, *Environ. Health Perspect.*, 62, 57, 1985.

19. **Randerath, K., Haglund, R.E., Phillips, D.H., and Reddy, M.V.,** ^{32}P-Postlabelling analysis of DNA adducts formed in the livers of animals treated with safrole, estragole and other naturally-occurring alkenylbenzenes. I. Adult female CD-1 mice, *Carcinogenesis*, 5, 1613, 1984.

20. **Phillips, D.H., Reddy, M.V., and Randerath, L.K.,** ^{32}P-Postlabelling analysis of DNA adducts formed in the livers of animals treated with saffrole, estragole and other naturally-occurring alkenylbenzenes. II. Newborn male B6C3F[1] mice, *Carcinogenesis*, 5, 1623, 1984.

21. **Warren, W.,** The analysis of alkylated DNA by high pressure liquid chromatography, in *Mutagenicity Testing: A Practical Approach*, Venitt, S. and Parry, J.M., Eds., IRL Press, Washington D.C., 1984, 25.

22. **Randerath, K., Reddy, M.V., and Gupta, R.C.,** ^{32}P-Labeling test for DNA damage, *Proc. Natl. Acad. Sci. U.S.A.*, 78, 6126, 1981.

23. **Gupta, R.C., Reddy, M.V., and Randerath, K.,** ^{32}P-Postlabeling analysis of non-radioactive aromatic carcinogen-DNA adducts, *Carcinogenesis*, 3, 1081, 1982.

24. **Watson, W.P.,** Post-radiolabelling for detecting DNA damage, *Mutagenesis*, 2, 319, 1987.

25. **Phillips, D.H.,** Modern methods of DNA adduct determination, in *Chemical Carcinogenesis and Mutagenesis I*, Cooper, C.S. and Grover, P.L., Eds., Springer-Verlag, Berlin, Heidelberg, New York, 1990, 503.

26. **Everson, R.B., Randerath, E., Santella, R.M., Cefalo, R.C., Avitts, T.A., and Randerath, K.,** Detection of smoking-related covalent DNA adducts in human placenta, *Science*, 231, 54, 1986.

27. **Pitot, H.C.,** Mechanisms of chemical carcinogenesis: theoretical and experimental bases, in *Chemical Carcinogenesis and Mutagenesis I*, Cooper, C.S. and Grover, P.L., Eds., Springer-Verlag, Berlin, Heidelberg, New York, 1990, 3.

28. **Purchase, I.H.F.,** An appraisal of predictive tests for carcinogenicity, *Mutat. Res.*, 99, 53, 1982.

29. **Tennant, R.W., Margolin, B.H., Shelby, M.D., Zeiger, E., Haseman, J.K., Spalding, J., Caspary, W., Resnick, M., Stasiewicz, S., Anderson, B., and Minor, R.,** Prediction of chemical carcinogenicity in rodents from in vitro genetic toxicity assays, *Science*, 236, 933, 1987.

30. **Zeiger, E., Haseman, J.K., Shelby, M.D., Margolin, B.H., and Tennant, R.W.,** A further evaluation of four *in vitro* genetic toxicity tests for predicting rodent carcinogenicity. Confirmation of earlier results with 41 additional chemicals, *Environ. Mol. Mutagen.*, *18* (Suppl.), 1, 1990.

31. **Brockman, H.E. and DeMarini, D.M.,** Utility of short-term tests for genetic toxicity in the aftermath of the NTP's analysis of 73 chemicals, *Environ. Mol. Mutagen.*, 11, 421, 1988.

32. **Flamm, W.G.,** A tier system approach to mutagen testing, *Mutat. Res.*, 26, 329, 1974.

33. **Weisburger, J.H. and Williams, G.M.,** Carcinogen testing: current problems and new approaches, *Science*, 214, 401, 1981.

34. **Williams, G.M. and Weisburger, J.H.,** Application of a cellular test battery in the decision point approach to carcinogen identification, *Mutat. Res.*, 205, 79, 1988.

35. **Cimino, M.C. and Auletta, A.E.,** The Gene-Tox Program: data evaluation of chemically-induced mutagenicity, in *Environmental Epidemiology, Advances in Chemistry Series No. 241*, American Chemical Society, Washington D.C., in press.

36. **Hansch, C. and Leo, A.,** *Substituent Constants for Correlation Analysis in Chemistry and Biology*, John Wiley & Sons, New York, 1979.

37. **Bandiera, S., Sawyer, T.W., Campbell, M.S., Fujita, T., and Safe, S.,** Competitive binding to the cytosolic 2,3,7,8-TCDD receptor: effects of structure on the affinities of substituted halogenated biphenyls — A QSAR approach, *Biochem. Pharmacol.*, 32, 3803, 1983.

38. **Denomme, M.A., Homonko, K., Fujita, T., Sawyer, T.W., and Safe, S.,** Effects of substituents on the cytosolic binding activities and AHH induction potencies of 7-substituted-2,3-dichlorodibenzo-*p*-dioxins — A QSAR analysis, *Mol. Pharmacol.*, 27, 656, 1985.

39. **Denomme, M.A., Homonko, K., Fujita, T., Sawyer, T.W., and Safe, S.,** Substituted polycholorinated dibenzofuran receptor binding affinities and aryl hydrocarbon hydroxylase induction potencies — a QSAR analysis, *Chem.-Biol. Interact.*, 57, 175, 1986.

40. **Enslein, K.,** Estimation of toxicological endpoints by structure-activity relationships, *Pharmacol. Rev.*, 36, 131S, 1984.

41. **Enslein, K., Blake, B.W., Tomb, M.E., and Borgstedt, H.H.,** Prediction of Ames test results by structure-activity relationships, *In Vitro Toxicol.*, 1, 33, 1986/1987.

42. **Enslein, K.,** An overview of structure-activity relationships as an alternative to testing in animals for carcinogenicity, mutagenicity, dermal and eye irritation, and acute oral toxicity, *Toxicol. Ind. Health*, 4, 479, 1988.

43. **Rosenkranz, H.S. and Klopman, G.,** CASE, the computer-automated structure evaluation system, as an alternative to extensive animal testing, *Toxicol. Ind. Health*, 4, 533, 1988.

44. **Rosenkranz, H.S. and Klopman, G.,** Structural basis of carcinogenicity in rodents of genotoxicants and nongenotoxicants, *Mutat. Res.*, 228, 105, 1990.

45. **Jerina, D.M., Lehr, R.E., Yagi, H., Hernandez, O., Dansette, P.M., Wislocki, P.G., Wood, A.W., Chang, R.L., Levin, W., and Conney, A.H.,** Mutagenicity of benzo[a]pyrene derivatives and the description of a quantum mechanical model which predicts the ease of carbonium ion formation from diol epoxides, in *In Vitro Metabolic Activation in Mutagenesis Testing*, de Serres, F.J., Fouts, J.R., Bend, J.R., and Philpot, R.M., Eds., Elsevier, Amsterdam, 1976, 159.

46. **Miller, E.C. and Miller, J.A.**, Mechanisms of chemical carcinogenesis: nature of proximate carcinogens and interactions with macromolecules, *Pharmacol. Rev.*, 18, 805, 1966.
47. **Miller, J.A.**, Carcinogenesis by chemicals: an overview, *Cancer Res.*, 30, 559, 1970.
48. **Heidelberger, C.**, Chemical carcinogenesis, *Annu. Rev. Biochem.*, 44, 79, 1975.
49. **Lijinsky, W.**, A view of the relation between carcinogenesis and mutagenesis, *Environ. Mol. Mutagen.*, 14 (Suppl. 16), 78, 1989.
50. **Schechtman, L.M.**, Metabolic activation of procarcinogens by subcellular enzyme fractions in the C3H 10T1/2 and BALB/c 3T3 cell transformation systems, in *Transformation Assay of Established Cell Lines: Mechanisms and Application*, Kakunaga, T. and Yamasaki, H., Eds., International Agency for Research on Cancer, Lyon, 1985, 143.
51. **Matsushima, T., Sawamura, M., Hara, K., and Sugimura, T.**, A safe substitute for polychlorinated biphenyls as an inducer of metabolic activation system, in *In Vitro Metabolic Activation in Mutagenesis Testing*, de Serres, F.J., Fouts, J.R., Bend, J.R., and Philpot, R.M., Eds., Elsevier/North-Holland, Amsterdam, 1976, 85.
52. **Maron, D.M. and Ames, B.N.**, Revised methods for the Salmonella mutagenicity test, *Mutat. Res.*, 113, 173, 1983.
53. INVITTOX Protocol No. 30, The Ames test, *The ERGATT/FRAME Data Bank of In Vitro Techniques in Toxicology*, 1992.
54. **Tamura, G., Gold, C., Ferro-Luzzi, A., and Ames, B.N.**, Fecalase: a model for activation of dietary glycosides to mutagens by intestinal flora, *Proc. Natl. Acad. Sci. U.S.A.*, 77, 4961, 1980.
55. **McCoy, E.C., Speck, W.T., and Rosenkranz, J.S.**, Activation of a procarcinogen to a mutagen by cell-free extracts of anaerobic bacteria, *Mutat. Res.*, 46, 261, 1977.
56. **Claxton, L.D., Allen, J., Auletta, A., Mortelmans, K., Nestmann, E., and Zeiger, E.**, Guide for the *Salmonella typhimurium*/mammalian microsome tests for bacterial mutagenicity, *Mutat. Res.*, 189, 83, 1987.
57. **Claxton, L.D., Houk, V.S., Allison, J.C., and Creason, J.**, Evaluating the relationship of metabolic activation system concentrations and chemical dose concentrations for the Salmonella spiral and plate assays, *Mutat. Res.*, 253, 127, 1991.
58. **Sugimura, T. and Nagao, M.**, Modification of mutagenic activity, in *Chemical Mutagens. Principles and Methods for their Detection*, Vol. 6, de Serres, F.J. and Hollaender, A., Eds., Plenum, New York, 1980, 41.
59. **Claxton, L.D., Houk, V.S., Warner, J.R., Myers, L.E., and Hughes, T.J.**, Assessing the use of known mutagens to calibrate the *Salmonella typhimurium* mutagenicity assay. II. With exogenous activation, *Mutat. Res.*, 253, 149, 1991.
60. **Langenbach, R. and Oglesby, L.**, The use of intact cellular activation systems in genetic toxicology assays, in *Chemical Mutagens. Principles and Methods for Their Detection*, Vol. 8., de Serres, F.J., Ed., Plenum, New York, 1983, 55.
61. **Bigger, C.A.H., Tomaszewski, J.E., and Dipple, A.**, Limitations of metabolic activation systems used with *in vitro* tests for carcinogens, *Science*, 209, 503, 1980.
62. **Bigger, C.A.H., Tomaszewski, J.E., and Dipple, A.**, Differences between products of binding of 7,12-dimethylbenz[a]anthracene to DNA in mouse skin and in a rat liver microsomal system, *Biochem. Biophys. Res. Commun.*, 80, 229, 1978.
63. **Madle, E., Tiedemann, G., Madle, S., Ott, A., and Kaufmann, G.**, Comparison of S9 mix and hepatocytes as external metabolizing systems in mammalian cell cultures: cytogenetic effects of 7,12-dimethylbenz[a]anthracene and aflatoxin B_1, *Environ. Mutagen.*, 8, 423, 1986.
64. **Autrop, H.**, Carcinogen metabolism in cultured human tissues and cells, *Carcinogenesis*, 11, 707, 1990.
65. **Donato, M.T., Gomez-Lechon, M.J., and Castell, J.V.**, Co-cultures of hepatocytes: *a biological model for long-lasting cultures*, in *In Vitro Alternatives to Animal Pharmacotoxicology*, Castell, J.V. and Gómez-Lechón, M.J., Eds., Farmaindustria, Madrid, 1992, 109.

66. **Mohn, G., Kerklaan, P., and Ellenberger, J.,** Methodologies for the direct and animal mediated determination of various genetic effects in derivatives of strain 343/113 of *Escherichia coli* K-12, in *Handbook of Mutagenicity Test Procedures,* 2nd ed., Kilbey, B.J., Legator, M., Nichols, W., and Ramel, C., Eds., Elsevier, Amsterdam, 1984, 189.

67. **Legator, M.S., Bueding, E., Batzinger, R., Connor, T.H., Eisenstadt, E., Farrow, M.G., Fiscor, G., Hsie, A., Seed, J., and Stafford, R.S.,** An evaluation of the host-mediated assay and bodyfluid analysis: a report of the U.S. Environmental Protection Agency Gene-Tox Program, *Mutat. Res.,* 98, 319, 1982.

68. **Connor, T.H. and Legator, M.S.,** The intraperitoneal host-mediated assay, in *Handbook of Mutagenicity Test Procedures,* 2nd ed., Kilbey, B.J., Legator, M., Nichols, W., and Ramel, C., Eds., Elsevier, Amsterdam, 1984, 643.

69. **Mirsalis, J.C. and Butterworth, B.E.,** Detection of unscheduled DNA synthesis in hepatocytes isolated from rats treated with genotoxic agents: an *in vivo-in vitro* assay for potential carcinogens and mutagens, *Carcinogenesis,* 1, 621, 1980.

70. **Butterworth, B.E., Ashby, J., Bermudez, E., Casciano, D., Mirsalis, J.C., Probst, G., and Williams, G.,** A protocol and guide for the *in vivo* rat hepatocyte DNA-repair assay, *Mutat. Res.,* 189, 123, 1987.

71. **Working, P.K. and Butterworth, B.E.,** An assay to detect chemically induced DNA repair in rat spermatocytes, *Environ. Mutagen.,* 6, 273, 1984.

72. **Ames, B.N.,** The detection of chemical mutagens with enteric bacteria, in *Chemical Mutagens: Principles and Methods for Their Detection,* Vol. 1, Hollaender, A., Ed., Plenum, New York, 1971, 267.

73. **McCann, J., Springarn, N.E., Kobori, J., and Ames, B.N.,** Detection of carcinogens as mutagens: bacterial tester strains with R factor plasmids, *Proc. Natl. Acad. Sci. U.S.A.,* 72, 979, 1975.

74. **Vicorin, K. and Stahlburg, M.,** A method for studying the mutagenicity of some gaseous compounds in *Salmonella Typhimurium, Environ. Mol. Mutagen.,* 11, 65, 1988.

75. **Callender, R.D.,** Use of mutations in bacteria as indicators of carcinogenic potential, in *Chemical Carcinogenesis and Mutagenesis II,* Cooper, C.S. and Grover, P.L., Eds., Springer-Verlag, Berlin, Heidelberg, New York, 1990, p. 3.

76. **Rosenkranz, H.S.,** Newer methods for determining genotoxicity using DNA repair-deficient and repair-proficient Escherichia coli, in *Handbook of Mutagenicity Test Procedures,* 2nd ed., Kilbey, B.J., Legator, M., Nichols, W., and Ramel, C., Eds., Elsevier, Amsterdam, 1984, 1.

77. **Leifer, Z., Kada, T., Mandel, M., Zeiger, E., Stafford, R., and Rosenkranz, H.S.,** An evaluation of tests using DNA repair-deficient bacteria for predicting genotoxicity and carcinogenicity: a report of the U.S. EPA's Gene-Tox Program, *Mutat. Res.,* 87, 211, 1981.

78. **Arlett, C.F.,** Mammalian cell mutations, in *Chemical Carcinogenesis and Mutagenesis II,* Cooper, C.S. and Grover, P.L., Eds., Springer-Verlag, Berlin, Heidelberg, New York, 1990, 27.

79. **Krahn, D.F.,** Chinese hamster cell mutagenesis: a comparison of the CHO and V79 systems, in *Cellular Systems for Toxicity Testing,* Williams, G.M., Dunkel, V.C. and Ray, V.A., Eds., New York Academy of Sciences, New York, 1983, 231.

80. **Langenbach, R., Hix, C., Oglesby, L., and Allen, J.,** Cell-mediated mutagenesis of Chinese hamster V79 cells and *Salmonella typhimurium,* in *Cellular Systems for Toxicity Testing,* Williams, G.M., Dunkel, V.C., and Ray, V.A., Eds., New York Academy of Sciences, New York, 1983, 258.

81. **Wild, D.,** Serum effect on the yield of chemically induced 8-azaguanine-resistant mutants in Chinese hamster cell cultures, *Mutat. Res.,* 25, 229, 1974.

82. **Peterson, A.R., Peterson, H., and Heidelberger, C.,** The influence of serum components on the growth and mutation of Chinese hamster cells in medium containing aminopterin, *Mutat. Res.,* 24, 25, 1974.

83. **Peterson, A.R., Krahn, D.F., Peterson, H., Heidelberger, C., Bhuyan, B.K., and Li, L.H.,** The influence of serum components on the growth and mutation of Chinese hamster cells in medium containing 8-azaguanine, *Mutat. Res.,* 36, 345, 1976.

84. **Subak-Sharpe, H., Burk, R.R., and Pitts, J.D.,** Metabolic cooperation between biochemically marked mammalian cells in tissue culture, *J. Cell Sci.,* 4, 353, 1969.

85. **Cox, R.P., Krauss, M.R., Balis, M.E., and Dancis, J.,** Evidence for transfer of enzyme product as the basis of metabolic cooperation between tissue culture fibroblasts of Lesch-Nyhan disease and normal cells, *Proc. Natl. Acad. Sci. U.S.A.,* 67, 1573, 1970.

86. **van Zeeland, A.A., van Diggelen, M.C.E., and Simons, J.W.I.M.,** The role of metabolic cooperation in selection of hypoxanthine-guanine-phosphoribosyl transferase (HGPRT)-deficient mutants from diploid mammalian cell strains, *Mutat. Res.,* 14, 355, 1972.

87. **Fox, M.,** Factors affecting the quantitation of dose-response curves for mutation induction in V-79 Chinese hamster after exposure to chemical and physical mutagens, *Mutat. Res.,* 29, 449, 1975.

88. **Yotti, L.P., Chang, C.-C., and Trosko, J.E.,** Elimination of metabolic cooperation in Chinese hamster cells by a tumor promoter, *Science,* 206, 1089, 1979.

89. **Elmore, E. and Nelmes, A.J.,** Summary report on the performance of the inhibition of metabolic cooperation assay, in *Progress in Mutation Research, Vol. 5,* Ashby, J., de Serres, F.J., Draper, M., Ishidate, M., Jr., Margolin, D.H., Matter, B.E., and Shelby, M.D., Eds., Elsevier, Amsterdam, 1985, 95.

90. **Swierenga, S.H.H. and Yamasaki, H.,** Performance of tests for cell transformation and gap-junction intercellular communication for detecting nongenotoxic carcinogenic activity, in *Mechanisms of Carcinogenesis in Risk Assessment,* Vainio, H., Magee, P.N., McGregor, D.B., and McMichael, A.J., Eds., International Agency for Research on Cancer (IARC), Lyon, 1992, 165.

91. **Loch-Caruso, R., Trosko, J.E., and Larson, W.J.,** Inhibited intercellular communication as a mechanistic link between teratogenesis and carcinogenesis, *Crit. Rev. Toxicol,* 16, 157, 1985.

92. **Hsie, A.W., Casciano, D.A., Couch, D.B., Krahn, D.F., O'Neill, J.P., and Whitfield, B.L.,** The use of Chinese hamster ovary cells to quantify specific locus mutation and to determine mutagenicity of chemicals: a report of the Gene-Tox Program, *Mutat. Res.,* 86, 193, 1981.

93. **Bradley, M.O., Bhuyan, B., Francis, M.C., Langenbach, R., Peterson, A., and Huberman, E.,** Mutagenesis by chemical agents in V79 Chinese hamster cells: a review and analysis of the literature. A report of the Gene-Tox Program, *Mutat. Res.,* 87, 81, 1981.

94. **Clive, D., McCuen, R., Spector, J.F.S., Piper, C., and Mavournin, K.H.,** Specific gene mutations in L5178Y cells in culture. A report of the U.S. Environmental Agency Gene-Tox Program, *Mutat. Res.,* 115, 225, 1983.

95. **Clive, D.,** Viable chromosomal mutations affecting the TK locus in L5178Y/TK$^{+/-}$ mouse lymphoma cells: the other half of the assay, in *Cellular Systems for Toxicity Testing,* Williams, G.M., Dunkel, V.C., and Ray, V.A., Eds., New York Academy of Sciences, New York, 1983, 253.

96. **Clive, D., Caspary, W., Kirby, P.E., Krehl, R., Moore, M., Mayo, J., and Oberly, T.J.,** Guide for performing the mouse lymphoma assay for mammalian cell mutagenicity, *Mutat. Res.,* 189, 143, 1987.

97. **Doerr, C.L., Harrington-Brock, K., and Moore, M.,** Micronucleus, chromosome aberration, and small-colony TK mutant analysis to quantitate chromosomal damage in L5178Y mouse lymphoma cells, *Mutat. Res.,* 222, 191, 1989.

98. **Jenssen, D.,** A quantitative test for mutagenicity in V79 Chinese hamster cells, in *Handbook of Mutagenicity Test Procedures,* 2nd ed., Kilbey, B.J., Legator, M., Nichols, W. and Ramel, C., Eds., Elsevier Amsterdam, 1984, 269.

99. **O'Neill, J.P., Brimer, P.A., Machanoff, R., Hirsch, G.P., and Hsie, A.W.,** A quantitative assay of mutation induction at the hypoxanthine-guanine phosphoribosyl transferase locus in Chinese hamster ovary cells (CHO/HGPRT system): development and definition of the system, *Mutat. Res.,* 45, 91, 1977.

100. **O'Neill, J.P., Couch, D.B., Machanoff, R., San Sebastian, J.R., Brimer, P.A., and Hsie, A.W.,** A quantitative assay of mutation induction at the hypoxanthine-guanine phosphoribosyl transferase locus in Chinese hamster ovary cells (CHO/HGPRT system): utilization with a variety of mutagenic agents, *Mutat. Res.,* 45, 103, 1977.

101. **Li, A.P., Carver, J.H., Choy, W.N., Hsie, A.W., Gupta, R.S., Loveday, K.S., O'Neill, J.P., Riddle, J.C., Stankowski, L.F., and Yang, L.L.,** A guide for the performance of the Chinese hamster ovary cell/hypoxanthine-guanine phosphoribosyl transferase gene muta- tion assay, *Mutat. Res.,* 189, 135, 1987.

102. **Li, A.P., Gupta, R.S., Helflich, R.H., and Wassom, J.S.,** A review and analysis of the Chinese hamster ovary/hypoxanthine guanine phosphoribosyl transferase assay to deter- mine the mutagenicity of chemical agents. A report of Phase III of the U.S. Environmental Protection Agency Gene-Tox Program, *Mutat. Res.,* 196, 17, 1988.

103. **Clive, D. and Spector, J.F.S.,** Laboratory procedure for assessing specific locus mutations at the TK locus in cultured L5178Y mouse lymphoma cells, *Mutat. Res.,* 31, 17, 1975

104. **Clive, D., Johnson, K.O., Spector, J.F.S., Batson, A.G., and Brown, M.M.M.,** Valida- tion and characterization of the L5178Y/TK$^{+/-}$ mouse lymphoma mutagen assay, *Mutat. Res.,* 59, 61, 1979.

105. **Turner, N.T., Batson, A.G., and Clive, D.,** Procedures for the L5178Y/TK$^{+/-}$ TK$^{-/-}$ mouse lymphoma cell mutagenicity assay, in *Handbook of Mutagenicity Test Procedures,* 2nd ed., Kilbey, B.J., Legator, M., Nichols, W., and Ramel, C., Eds., Elsevier, Amsterdam, 1984, 239.

106. **Nestmann, E.R., Brillinger, R.L., Gilman, J.P.W., Rudd, C.J., and Swierenga, S.H.H.,** Recommended protocols based on a survey of current practice in genotoxicity testing laboratories. II. Mutation in Chinese hamster ovary, V79 Chinese hamster lung and L5178Y mouse lymphoma cells, *Mutat. Res.,* 246, 255, 1991.

107. **Lambert, B., Chu, E.H.Y., De Carli, L., Ehling, U.H., Evans, H.J., Hayashi, M., Thilly, W.G., and Vainio, H.,** Assays for genetic changes in mammalian cells, Report 7, in *Long- Term and Short-Term Assays for Carcinogens,* Montesanto, R., Bartsch, H., Vainio, H., Wilbourn, J., and Yamasaki, H., Eds., International Agency for Research in Cancer, Lyon, 1986, 167.

108. **Preston, R.J., Au, W., Bender, M.A., Brewen, J.G., Carrano, A.V., Heddle, J.A., McFee, A.F., Wolff, S., and Wassom, J.S.,** Mammalian *in vivo* and *in vitro* cytogenetic assays: a report of the U.S. Environmental Protection Agency Gene-Tox Program, *Mutat. Res.,* 87, 143, 1981.

109. **Ishidate Jr., M.,** *Data Book of Chromosomal Aberration Test In Vitro,* rev. ed., Elsevier, Amsterdam, 1988, 15.

110. **Taylor, J.H.,** Sister chromatid exchanges in tritium labelled chromosomes, *Genetics,* 43, 515, 1958.

111. **Perry, P.E. and Wolff, S.,** New Giemsa method for the differential staining of sister chromatids, *Nature,* 251, 156, 1974.

112. **Perry, P.E.,** Chemical mutagens and sister-chromatid exchange, in Chemical Mutagens, Vol. 6, de Serres, F.J. and Hollaender, A., Eds., Plenum, New York, 1980, 1.

113. **Carrano, A.V.,** Sister chromatid exchange as an indicator of human exposure, in *Banbury Report No. 13,* Cold Spring Harbour, New York, 1982, 307.

114. **Heddle, J.A.,** A rapid *in vivo* test for chromosomal damage, *Mutat. Res.,* 18, 187, 1973.

115. **Heddle, J.A., Hite, M., Kirkhart, B., Mavournin, K., Macgregor, J.T., Newell, G.W., and Salamone, M.F.,** The induction of micronuclei as a measure of genotoxicity. A report of the U.S. Environmental Protection Agency Gene-Tox Program, *Mutat. Res,* 123, 61, 1983.

116. **Heddle, J.A., Stuart, E., and Salamone, M.F.,** The bone marrow micronucleus test, in *Handbook of Mutagenicity Testing,* 2nd ed., Kilbey, B.J., Legator, M., Nichols, W. and Ramel, C., Eds., Elsevier, Amsterdam, 1984, 442.

117. **Fenech, M. and Morley, A.A.,** Measurement of micronuclei in lymphocytes, *Mutat. Res.,* 147, 29, 1985.

118. **Swierenga, S.H.H, Heddle, J.A., Sigal, E.A., Gilman, J.P.W., Brillinger, R.L., Douglas, G.R., and Nestmann, E.R.,** Recommended protocols based on a survey of current practice in genotoxicity testing laboratories. IV. Chromosome aberration and sister-chromatid exchange in Chinese hamster ovary, V79 Chinese hamster lung and human lymphocytes, *Mutat. Res.,* 246, 301, 1991.

119. **Wolff, S.,** Sister chromatid exchanges as a test for mutagenic carcinogens, in *Cellular Systems for Toxicity Testing,* Williams, G.M., Dunkel, V.C., and Ray V.A., Eds., The New York Academy of Sciences, New York, 1983, 142.

120. **Perry, P.E. and Thomson, E.F.,** The methodology of sister chromatid exchanges, in *Handbook of Mutagenicity Test Protocols,* 2nd ed., Kilbey, B.J., Legator, M., Nichols, W., and Ramel, C., Eds., Elsevier, Amsterdam, 1984, 496.

121. **Latt, S.A., Allen, J., Bloom, S.E., Carrano, A., Falke, E., Kram, D., Schneider, E., Schreck, R., Tice, R., Whitfield, B., and Wolff, S.,** Sister-chromatid exchanges: a report of the Gene-Tox Program, *Mutat. Res.,* 87, 17, 1981.

122. **Wolff, S.,** Sister chromatid exchange, *Annu. Rev. Genet.,* 11, 183, 1977.

123. **Roberts, J.J.,** The repair of DNA modified by cytotoxic, mutagenic and carcinogenic chemicals, *Adv. Radiat. Biol.,* 7, 211, 1978.

124. **Williams, G.M.,** Further improvements in the hepatocyte primary culture DNA repair test for carcinogens: detection of carcinogenic biphenyl derivatives, *Cancer Lett.,* 4, 69, 1978.

125. **Althaus, F.R., Lawrence, S.D., Sattler, G.L., Longfellow, D.G., and Pitot, H.C.,** Chemical quantification of unscheduled DNA synthesis in cultured hepatocytes as an assay for the rapid screening of potential chemical carcinogens, *Cancer Res.,* 42, 3010, 1982.

126. INVITTOX Protocol No. 18, Unscheduled DNA synthesis in hepatocyte cultures assessed by the nuclei procedure, in *The ERGATT/FRAME Data Bank of In Vitro Techniques in Toxicology,* 1991.

127. **Hsia, M.T.S., Kreamer, B.L., and Dolara, P.,** A rapid and simple method to quantitate chemically induced unscheduled DNA synthesis in freshly isolated rat hepatocytes facilitated by DNA retention of membrane filters, *Mutat. Res.,* 122, 177, 1983.

128. **Bohr, V., Mansbridge, J., and Hanawalt, P.,** Comparative effects of growth inhibitors on DNA replication, DNA repair, and protein synthesis in human epidermal keratinocytes, *Cancer Res.,* 46, 2929, 1986.

129. **Rasmussen, R.E. and Painter, R.B.,** Evidence for repair of ultra-violet damaged deoxyribonucleic acid in cultured mammalian cells, *Nature,* 203, 1360, 1964.

130. **Rasmussen, R.E. and Painter, R.B.,** Radiation-stimulated DNA synthesis in cultured mammalian cells, *J. Cell Biol.,* 29, 11, 1966.

131. **Stich, H.F., San, R.H.C., and Kawazoe, Y.,** DNA repair synthesis in mammalian cells exposed to a series of oncogenic and non-oncogenic derivatives of 4-nitroquinoline-1-oxide, *Nature,* 299, 416, 1971.

132. **Williams, G.M.,** Carcinogen-induced DNA repair in primary rat liver cell cultures: a possible screen for chemical carcinogens, *Cancer Lett.,* 1, 231, 1976.

133. **Williams, G.M.,** Detection of chemical carcinogens by unscheduled DNA synthesis in rat liver primary cell cultures, *Cancer Res.,* 37, 1845, 1977.

134. **Butterworth, B.E.,** Measurement of chemically induced DNA repair in rodent and human cells, in *New Approaches in Toxicity Testing and Their Application to Human Risk Assessment,* Li, A.P., Ed., Raven Press, New York, 1983, 33.

135. **Mitchell, A.D. and Mirsalis, J.C.,** Unscheduled DNA synthesis as an indicator of genotoxic exposure, in *Single-Cell Mutation Monitoring Systems,* Ansari, A.A. and de Serres, F.J., Eds., Plenum, New York, 1984, 165.

136. **McQueen, C.A.,** Hepatocytes in monolayer culture: An *in vitro* model for toxicity studies, in *In Vitro Toxicology: Model Systems and Methods,* McQueen, C.A., Ed., Telford Press, Caldwell, NJ, 1989, 131.

137. **Mirsalis, J.C., Tyson, C.K., Loh, E.N., Steinmetz, K.L., Bakke, J.P., Hamilton, C.M., Spak, D.K., and Spalding, J.W.,** Induction of hepatic cell proliferation and unscheduled DNA synthesis in mouse hepatocytes following *in vivo* treatment, *Carcinogenesis,* 6, 1521, 1985.

138. **International Agency for Research on Cancer,** Long-term and short-term screening assays for carcinogens: a critical appraisal, *Int. Agen. Res. Cancer, Monogr.,* Suppl. 2, Lyon, 1980.

139. **Larsen, K.H., Brash, D., Cleaver, J.E., Hart, R.W., Maher, V.M., Painter, R.B., and Sega, G.A.,** DNA repair assays as tests for environmental mutagens. A report of the U.S. Environmental Protection Agency Gene-Tox Program, *Mutat. Res.,* 98, 287, 1982.

140. **Williams, G.M., Mori, H., and McQueen, C.A.,** Structure-activity relationships in the rat hepatocyte DNA-repair test for 300 chemicals, *Mutat. Res.,* 221, 263, 1989.

141. **Fautz, R., Forster, R., Hechenberger, C.M.A., Hertner, T., von der Hude, W., Kaufman, G., Madle, H., Madle, S., Miltenberger, H.G., Muller, L., Pool-Zobel, B.L., Puri, E.C., Schmezer, P., Seeberg, A.H., Strobel, R., Suter, W., and Baumeister, M.,** Report of a comparative study of DNA damage and repair assays in primary rat hepatocytes with five coded chemicals, *Mutat. Res.,* 260, 281, 1991.

142. **Mitchell, A.D., Casciano, D.A., Meltz, M.L., Robinson, D.E., San, R.H.C., Williams, G.M., and Von Halle, E.S.,** Unscheduled DNA synthesis tests. A report of the U.S. Environmental Protection Agency Gene-Tox Program, *Mutat. Res.,* 123, 363, 1983.

143. **Butterworth, B.E., Ashby, J., Bermudez, E., Casciano, D., Mirsalis, J., Probst, G., and Williams, G.M.,** A protocol and guide for the in vitro rat hepatocyte DNA-repair assay, *Mutat. Res.,* 189, 113, 1987.

144. **Butterworth, B.E.,** Measurement of chemically induced DNA repair in hepatocytes *in vivo* and *in vitro* as an indicator of carcinogenic potential, in *The Isolated Hepatocyte. Use in Toxicology and Xenobiotic Biotransformation,* Rauckman, E.J. and Padilla, G.M., Eds., Academic Press, New York, 1987, 241.

145. **Swierenga, S.H.H., Bradlaw, J.A., Brillinger, R.L., Gilman, J.P.W., Nestmann, E.R., and San, R.C.,** Recommended protocols based on a survey of current practice in genotoxicity testing laboratories. I. Unscheduled DNA synthesis assay in rat hepatocyte cultures, *Mutat. Res.* 246, 235, 1991.

146. **Pienta, R.J., Poiley, J.A., and Raineri, R.,** Application of transformation systems, in *Cellular Systems for Toxicity Testing,* Williams, G.M., Dunkel, V.C., and Ray, V.A., Eds., New York Academy of Sciences, New York, 1983, 267.

147. **McGregor, D. and Ashby, J.,** Summary report on the performance of the cell transformation assays, in *Progress in Mutation Research.* Vol. 5, Ashby, J., de Serres, F.J., Draper, M., Ishidate Jr., M., Margolin, B.H., Matter, B.E., and Shelby, M.D., Eds., Elsevier, Amsterdam, 1985, 103.

148. **Earle, W.R. and Nettleship, A.,** Production of malignancy in vitro. V. Results of injections of cultures into mice, *J. Natl. Cancer Inst.,* 4, 213, 1943.

149. **Berwald, Y. and Sachs, L.,** *In vitro* transformation with chemical carcinogens, *Nature,* 200, 1182, 1963.

150. **Berwald, Y. and Sachs, L.,** *In vitro* transformation of normal cells to tumor cells by carcinogenic hydrocarbons, *J. Natl. Cancer Inst.,* 35, 641, 1965.

151. **Styles, J.A.,** A method for detecting carcinogenic organic chemicals using mammalian cells in culture, *Br. J. Cancer,* 36, 558, 1977.

152. **Pienta, R.J., Poiley, J.A., and Lebherz III, W.B.,** Morphological transformation of early passage golden Syrian hamster embryo cells derived from cryopreserved primary cultures as a reliable *in vitro* bioassay for identifying diverse carcinogens, *Int. J. Cancer,* 19, 642, 1977.

153. **Kakunaga, T.,** A quantitative system for assay of malignant transformation by chemical carcinogens using a clone derived from Balb/3T3, *Int. J. Cancer,* 12, 463, 1973.

154. **Reznikoff, C.A., Bertram, J.S., Brankow, D.W., and Heidelberger, C.,** Quantitative and qualitative studies of chemical transformation of cloned C3H mouse embryo cells sensitive to post-confluence inhibition of cell division, *Cancer Res.,* 33, 3239, 1973.

155. **Reznikoff, C.A., Brankow, D.W., and Heidelberger, C.,** Establishment and characterization of a cloned line of C3H embryo cells sensitive to post-confluence inhibition of division, *Cancer Res.,* 33, 3231, 1977.

156. **Styles, J.A.,** Mammalian cell transformation *in vitro, Br. J. Cancer,* 37, 931, 1977.

157. **Purchase, I.F.H., Longstaff, E., Ashby, J., Styles, J.A., Anderson, D., Lefevre, P.A., and Westwood, F.R.,** Evaluation of six short-term tests for detecting organic chemical carcinogens and recommendations for their use, *Nature,* 264, 624, 1976.

158. **Purchase, I.F.H., Longstaff, E., Ashby, J., Styles, J.A., Anderson, D., Lefevre, P.A., and Westwood, F.R.,** An evaluation of six short-term tests for detecting organic chemical carcinogens, *Br. J. Cancer,* 37, 873, 1978.

159. **Meyer, A.L.,** In vitro transformation assays for chemical carcinogens, *Mutat. Res.,* 115, 323, 1983.

160. **de Serres, F.J. and Ashby, J., Eds.,** *Short Term Tests for Carcinogenicity: Report of the International Collaborative Program,* Elsevier/North Holland, Amsterdam, 1981.

161. **Dunkel, V.C., Rogers, C., Swierenga, S.H.H., Brillinger, R.L., Gilman, J.P.W., and Nestmann, E.R.,** Recommended protocols based on a survey of current practice in genotoxicity testing laboratories. III. Cell transformation in C3H/10T1/2 mouse embryo cell, BALB/c 3T3 mouse fibroblast and Syrian hamster embryo cell cultures, *Mutat. Res.,* 246, 285, 1991.

162. **Mondal, S., Brankow, D.W., and Heidelberger, C.,** Two-stage chemical oncogenesis in cultures of C3H/10 1/2 cells, *Cancer Res.,* 36, 2254, 1976.

163. **Mondal, S. and Heidelberger, C.,** Transformation of C3H/10T 1/2 Cl 8 mouse embryo fibroblasts by ultraviolet irradiation and a phorbol ester, *Nature,* 260, 710, 1976.

164. **Sakai, A. and Sato, M.,** Improvement of carcinogen identification in BALB/3T3 cell transformation by application of a 2-stage method, *Mutat. Res.,* 214, 285, 1989.

165. **DiPaolo, J.A., Donovan, P.J., and Nelson, R.L.,** Quantitative studies of *in vitro* transformation by chemical carcinogens, *J. Natl. Cancer Inst.,* 42, 867, 1969.

166. **DiPaolo, J.A., Nelson, R.L., and Donovan, P.J.,** *In vitro* transformation of Syrian hamster embryo cells by diverse chemical carcinogens, *Nature,* 235, 278, 1972.

167. **DiPaolo, J.A., Nelson, R.L., and Donovan, P.J.,** Oncogenic and karyological characteristics of Syrian hamster embryo cells transformed *in vitro* by carcinogenic polycyclic hydrocarbons, *Cancer Res.,* 31, 1118, 1971.

168. **Casto, B.C., Janosko, N., and DiPaolo, J.A.,** Development of a focus assay model for transformation of hamster cells *in vitro* by chemical carcinogens, *Cancer Res.,* 37, 3508, 1977.

169. **Sivak, A. and Tu, A.S.,** Use of rodent hepatocytes for metabolic activation in transformation assays in *Transformation Assay of Established Cell Lines: Mechanisms and Application,* Kagunaga, T. and Yamasaki, H., Eds., International Agency for Research on Cancer, Lyon, 1985, 121

170. **Casto, B.C., Pieczynski, W.J., and DiPaolo, J.A.,** Enhancement of adenovirus transformation by pretreatment of hamster cells with carcinogenic polycyclic hydrocarbons, *Cancer Res.,* 33, 819, 1973.

171. **Casto, B.C., Pieczynski, W.J., and Dipaolo, J.A.,** Enhancement of adenovirus transformation by treatment of hamster embryo cells with diverse chemical carcinogens, *Cancer Res.,* 34, 72, 1974.

172. **Freeman, A.E., Price, P.J., Igel, H.G., Young, J.C., Maryak, J.M., and Huebner, R.L.,** Morphological transformation of rat embryo cells induced by diethylnitrosamine and murine leukemia viruses, *J. Natl. Cancer Inst.,* 44, 65, 1970.

173. **Tu, A., Hallowell, W., Pallotta, S., Sivak, A., Lubet, R.A., Curren, R.D., Avery, M.D., Jones, C., Sedita, B.A., Huberman, E., Tennant, R., Spalding, J., and Kouri, R.E.,** An interlaboratory comparison of transformation in Syrian hamster embryo cells with model and coded chemicals, *Environ. Mutagen.*, 8, 77, 1986.

174. **Dunkel, V.C., Schechtman, L.M., Tu, A.S., Sivak, A., Lubet, R.A., and Cameron, T.P.,** Interlaboratory evaluation of the C3H/10T1/2 cell transformation assay, *Environ. Mol. Mutagen.*, 12, 21, 1988.

175. **Heidelberger, C., Freeman, A.E., Pienta, R.J., Sivak, A., Bertram, J.S., Casto, B.C., Dunkel, V.C., Francis, M.W., Kakunaga, T., Little, J.B., and Schechtman, L.M.,** Cell transformation by chemical agents — a review and analysis of the literature. A report of the U.S. Environmental Protection Agency Gene-Tox Program, *Mutat. Res.*, 114, 283, 1983.

176. **Kakunaga, T.,** Critical review of the use of established cell lines for *in vitro* cell transformation, in *Transformation Assay of Established Cell Lines: Mechanisms and Applications,* Kakunaga, T. and Yamasaki, H., Eds., International Agency for Research on Cancer, Lyon, 1985, 55.

177. **Schechtman, L.M.,** BALB/c 3T3 cell transformation: Protocols, problems and improvements, in *Transformation Assay of Established Cell Lines: Mechanism and Application,* Kakunaga, T. and Yamasaki, H., Eds., International Agency for Research in Cancer, Lyon, 1985, 165.

178. **Landolph, J.R.,** Chemical transformation in C3H 10T1/2 Cl 8 mouse embryo fibroblasts: historical background, assessment of the transformation assay, and evolution and optimization of the transformation assay protocol, in *Transformation Assay of Established Cell Lines: Mechanisms and Application,* Kakunaga, T. and Yamasaki, H., Eds., International Agency for Research on Cancer, Lyon, 1985, 185.

179. **Working Group,** Recommendations for experimental protocols and for scoring transformed foci in BALB/c 3T3 and C3H 10T1/2 cell transformation, in *Transformation Assay of Established Cell Lines: Mechanisms and Application,* Kagunaga, T. and Yamasaki, H., Eds., International Agency for Research in Cancer, Lyon, 1985, 207.

Chapter 10

EXPERIMENTAL DESIGN AND STATISTICS

I. PRACTICAL CONSIDERATIONS

A. EXPERIMENTAL SETUP

Of fundamental importance in the setup of an experiment is the choice of the materials and suppliers. In general, the manufacturer or distributor of the materials and chemicals, including the supplies for cell culture, should be reputable, easily accessible, and referenced. In addition, the company should provide the researcher with details of the methods used to prepare, package, store, and ship supplies. It is also important that the research laboratory maintain records of dates and storage of chemicals, and the lab should be consistent in its choice of company. For instance, it would be ideal if one company could provide the majority of compounds to be tested. Among the materials which should be obtained include cell culture supplies, stock cell cultures, chemicals, isotopes, and materials for scintillation counting. In addition, animal farms should provide the investigator with the circumstances of breeding and housing, as well as a physical examination of the animals and a declaration of whether they are certified as pathogen-free.

During the manipulations of the cells in culture, certain details should be established and followed judiciously and uniformly. The rate of inoculating cells depends on the size of the flask or wells. For instance, 10-, 25-, 75-, and 150-cm^2 flasks are suitable for individual experiments or for maintaining cells in continuous passage. Also, 6-, 12-, 24-, 48-, and 96-well plates, corresponding to decreasing surface area per well, are routinely employed in screening assays. In general, cells are usually seeded at 10^4 cells per square centimeter. Contact inhibited cultures are grown to confluency, unless cell growth experiments are performed, and subcultured in appropriate complete medium supplemented with fetal bovine serum and antibiotics, if needed. Alternatively, serum-free medium is used in a consistent manner. Depending on the doubling rate of the cell line, the time required for confluent monolayers to grow can be predicted with accuracy. For example, when seeded at one-third the confluent density, or according to the formula above, most continuous cell lines with appreciable doubling rates reach the stationary phase within 4 to 8 days.

Monolayers of cells are then incubated with increasing concentrations of each chemical in at least 4 doses (3 to 4 wells per dose) for a predetermined period of time at 37°C in a gaseous atmosphere which is defined by the requirements of the medium. This atmosphere may consist of 5 to 20% carbon dioxide in air, unless otherwise stated. A stock solution of the test chemical is prepared in the incubating medium and serial dilutions are made from the stock solution. This method of formulating a stock solution decreases the chances of introducing an error when adding chemical to each experimental group. In addition, for experiments requiring longer incubation times, the media may be sterilized initially, so that aseptic conditions are more easily maintained. A soluble solution of the chemical is sterilized with submicron filters, while insoluble solids are first exposed to ultraviolet light for several minutes in an open test tube before addition of buffer. In a time-response experiment, the concentration for all groups remains the same, but the groups are terminated at preset, incremental time periods. Before the experiment is terminated, indicators, dyes, fixatives, or reactive substances are added as needed and allowed to incubate. The cells are then counted or processed, and the reaction product is quantified according to the protocol.

In radiolabeling studies, a radioactive (labeled) precursor is added to the cultures, usually 1 hour after the medium has been changed from the growth medium to the labeling medium. The precursor can be an amino acid for protein synthesis, a monosaccharide for carbohydrate metabolism, a precursor for lipid biosynthesis, or any substance designed to *trace* the intracellular biosynthetic pathway of a molecule. The labeling medium should be a simpler version of the growth medium. Care is taken to avoid the presence of the precursor, or a parent molecule, in this formula. In addition, the serum, generally used during the growth stage, should be dialyzed. Other supplements to the incubation medium may include buffers, antibiotics, or cofactors necessary for the proper synthesis of the molecules being measured. Cell viability is usually measured in separate cultures and is expressed as a percentage of the control groups. Additional wells without cells are processed with the appropriate media in parallel as reference blanks.

B. INCUBATION MEDIUM AND TEST
CHEMICAL INCOMPATIBILITY

Most of the technical problems involving the setup and execution of toxicity testing experiments involve the solubilization or miscibility of the chemicals with the incubation medium. Solid chemicals that are insoluble in water, such as paracetamol (acetaminophen), and acetylsalicylic acid (aspirin), present with dissolution problems, especially at higher dosage levels. The solubility of these chemicals may be improved by a variety of techniques, as shown in Table 1.

1. Micronization

This procedure increases the surface area of a solid and can improve the solubility of a powder at low doses. A solution of the chemical is prepared in ethanol or dimethylsulfoxide (DMSO) for each group, and the solution is evaporated to dryness under a stream of nitrogen gas. The remaining lyophilized powder is then dissolved more readily in medium using constant stirring at 37°C for at least 1 h prior to incubation.

TABLE 1
Outline of Methods Used to Increase Solubility and Miscibility of Chemicals

Manipulation	Form of chemical	Method
Micronization	Solid dissolved in alcohol or DMSO[a]	Evaporation of stock alcoholic solution of chemical, followed by constant stirring of lyophilized powder at 37°C × 1 h
Stock solution in ethanol or DMSO	Solid dissolved in alcohol or DMSO	Concentrated stock solution of test chemical is formulated and aliquots are added to media of test groups as required
Sonication	Organic liquid	Appropriate quantity of immiscible organic liquid is added to medium and sonicated with an ultrasonic probe (30 W × 10 s.)
Mineral oil overlay	Organic liquid	Appropriate quantity of volatile test substance is added to medium, and mineral oil is layered over the surface; flasks are gassed and sealed individually

[a]   Dimethylsulfoxide.

2. Use of Solvents to Improve Solubility

At other times, it may be necessary to dissolve the test chemical in a *solvent* which is miscible with the incubation medium, such as ethanol or DMSO. A stock solution of the substance is prepared in the solvent and appropriate aliquots are then distributed to the corresponding experimental groups. It is important, however, that individual laboratories determine the toxicity of the solvent, even at its lowest concentration as part of the incubation media. Also, all experimental groups should be equilibrated with the same amount of solvent, including the controls. This will negate any minor effects of the solvent alone on the cells.

3. Sonication

Other organic liquids used as test chemicals, such as xylene or carbon tetrachloride, are immiscible with water and culture media. Miscibility is improved by *sonicating* a stock solution of the chemical in medium. This usually requires either a hand-held teflon pestle and glass tube so that the combined liquids can be hand homogenized, or, alternatively, and much more effectively, an ultrasonic processor equipped with a longer probe for culture tubes, is inserted in the immiscible liquids for as little as 10 seconds at 10 to

50 watts of power. This manipulation completely homogenizes the mixture and allows enough time for adequately dispersing an aliquot of the mixture into the incubation medium. On occasion, the liquids separate out during the incubation time. In this case, the mixture is resuspended and added again to the cultures.

4. Use of Paraffin (Mineral) Oil Overlay

More volatile chemicals, such as the alcohols or chloroform, are especially annoying, since they permeate the incubator atmosphere (and sometimes the laboratory) and interfere with control wells. The evaporation, and thus the concentration of the chemical in the medium, can be controlled by incubating the chemicals with cells previously grown in 25 cm^2 (T-25) flasks, which are then overlaid with a thin layer of paraffin oil (light mineral oil). The oil suppresses the evaporation of volatile molecules at the surface by increasing the surface tension. The flasks, however, must be individually permeated with a stream of gas corresponding to the gaseous atmosphere of the incubator and sealed with screw caps. This requires some forethought, since it is necessary to anticipate which chemical will be tested and which vessels will be used to grow the cells. A separate gas tank, formulated for the incubation medium, is also needed for these studies.

The presence of an oil overlay may produce enough surface tension so as to prevent evaporation of volatile chemicals partially dissolved in the medium. The partial pressure of the chemical in a liquid, however, may determine to what extent the test substance partitions between the air and gas phases. Ultimately this partitioning will determine the amount of substance dissolved in the incubating medium and in contact with the cell layer. Essentially, it may be necessary to determine the actual concentration of the chemical in the liquid and the air space above it in a series of control flasks. The method of choice may include ***head-space gas chromatography***. In this way an accurate determination of the kinetics of air-liquid distribution is determined.

C. DETERMINATION OF DOSAGE RANGE

It is not always necessary to test logarithmic concentrations of a chemical, especially when resources are limited. If the agent has been tested in animals and the LD$_{50}$ is known, this value can be used to estimate the human equivalent toxic concentration in plasma.[1,2] The initial concentrations of a chemical used for the *in vitro* experiments, therefore, are estimated from the known human toxicity data or is derived from rodent LD$_{50}$ values[3-5] according to the following formula:

$$\text{HETC} = \left(\text{LD}_{50}\right)/\text{V}_\text{d} * 10^{-3}$$

where HETC = estimated human equivalent toxic concentration in plasma (mg/ml); LD$_{50}$ = 50% lethal dose in rodents (mg/kg, intraperitoneal or oral); V$_\text{d}$ = volume of distribution (L/kg); and, 10^{-3} = constant for conversion into ml(L/ml).

This value is an estimate of the toxic human blood concentration equivalent to the LD_{50} dose given to a rodent, based on body weight. The HETC value is then used as a guideline for establishing the dosage range for each group of an experiment. HETC values can also be used as a method of converting animal LD_{50} data into *equivalent* human toxicity information, which is valuable as a procedure for comparing *in vitro* IC_{50}s against human lethal concentrations derived from the clinical case studies and poison control centers. Essentially, the formula offers a mechanism for converting rodent *in vivo* data to equivalent human toxicity guidelines, in order to compare with *in vitro* data in their ability to predict human toxicity.

D. DETERMINATION OF INHIBITORY CONCENTRATIONS

Figure 1 shows typical concentration-effect curves generated for representative chemicals using [^{3}H]proline as an indicator for protein synthesis.[6,7] The concentrations necessary to inhibit 50% (IC_{50}) of the newly synthesized proteins is then calculated based on the slope and linearity of the curve. The symmetry of the curve is mathematically estimated from the ***regression analysis*** (see Section II, Statistical Methods) of the plot, "% of control" vs. "log of concentration". The values for "% of control" are derived by converting the absolute values of the measured response to a fraction of the control value; that is, the measured value of the group with no chemical. Using the control value as the 100% level, all subsequent groups are compared to this. Thus, absolute values are transformed into relative percentage values. This has the added advantage that different experiments can be compared, even when the absolute values of the control groups are not identical.

When the response is plotted against the log of the concentration used for that group, a straight line of "best fit" is obtained. The IC_{50} is then estimated by extrapolating from the 50% response horizontally until it intersects the plot, and vertically from that point until it reaches a value on the X-axis. Of course, to obtain a line of best fit, regression analysis must be used for the plot. Today, relatively inexpensive calculators can be summoned to perform this calculation, if the student does not have access to a statistical computer program.

The plot for some representative substances, grouped according to chemical classification, in Figure 1 shows that the regression lines are significantly accurate so that the IC_{50}s can be computed with statistical confidence. (The plots for these chemicals have calculated correlation coefficients [r values] between 0.90 and 0.99.) In a similar manner, the 10 and 75% inhibitory concentrations (IC_{10}s and IC_{75}s, respectively) are also computed to demonstrate the lower and upper cutoff points. In addition, the cutoff points aid in the determination of the range of toxicity; i.e., the greater the difference between the upper and lower ICs, the wider the range of toxicity for a particular chemical. A chemical possessing a smaller difference between the IC_{10} and IC_{75} has a narrower range of toxicity. This implies that there is fine margin between a nontoxic and toxic dose, or between toxicity and lethality. When the

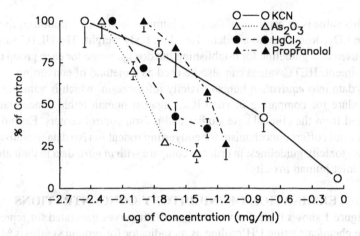

Figure 1. Concentration-effect curves for representative chemicals in cultured rat lung cells using the 24-h [³H]proline assay comparing relative *in vitro* cytotoxicities. Inhibitory concentrations at the 50% level were determined from these plots.

IC_{50} is not bracketed in the dosage range used for the chemical, the experiment is repeated with different sets of concentrations.

II. STATISTICAL METHODS

This section is not designed to explain the mechanics of statistical manipulations or to propose an exhaustive review of statistics for toxicologists. Moreover, it assumes that the student and investigator understand some basic principles of statistics, since several introductory areas are not broached. These areas include descriptive statistics (mean, median, standard deviation); types of variables (continuous and discontinuous data); normal, bimodal, and Poisson frequency distribution; and parametric and nonparametric statistics. These and other terms, however, are defined in Table 2. Consequently, for the mathematical manipulations, the reader is referred to some of the excellent reference texts currently available in the application of statistics in toxicology.

The purpose of this chapter, therefore, is to inform the student of the different mathematical tests used most often for analyzing a set of data. Some forethought is required in the experimental setup so that the appropriate statistical approach can be employed. These calculations are incorporated into a study either alone or combined with other statistical parameters, thus reinforcing the validity of the experiments. In addition, the type and number of statistical tests employed may reduce the chances that a nonsignificant result will be declared significant, especially when individual experiments are repeated many times. It is an innate consequence, however, that the more often the same statistical test is repeated for each experiment, the greater are the chances of rejecting the true null hypothesis.

TABLE 2
Definitions of Some Fundamental Terms in Statistics for Interpreting
In Vitro Cytotoxicity Testing Data

Independent variable — a characteristic or quantifiable parameter that can be assigned different values within the test system; often referred to as **random** variables when the values are obtained as a result of chance factors; example includes established concentration of a chemical.

Dependent variable — values resulting from measurement procedures and relying on the value of the independent variable; referred to as **observations** or **measurements**, including quantitation of cell growth or level of enzyme activity after treatment with a chemical.

Continuous data — data which can be plotted as points on a line, including any quantifiable parameter, but which can assume any of an infinite number of values between two fixed points, such as body weight or protein concentration.

Discontinuous data — quantifiable parameters as with continuous data but which have only fixed values, with no possible intermediate values, such as number of cells alive or dead.

Probability — the frequency with which a particular value will be found, given an appropriately sized sample.

Frequency distribution — a table or graph in which the values of the variable are distributed among the specified class intervals (that is, a set of nonoverlapping observations).

Hypothesis testing — a determination of whether or not a statement about one or more populations is compatible with available data; the hypothesis is concerned with the parameters of the populations about which the statement is made, and testing aids the investigator in reaching a decision about a population by examining a sample from that population.

Significance level — all possible values that the test statistic can assume are graphed as points on the horizontal axis of the distribution and are divided into an **acceptance region** and a **rejection region**. The decision as to which of the experimental values (observations) go into the regions is made according to the selected level of significance, indicated as α. The level specifies the area under the curve of distribution of the test statistic that is above the values on the horizontal axis constituting the rejection region. The significance level is often set at 0.001, 0.01 or 0.05 corresponding to the 99.9%, 99.0% and 95% confidence intervals, respectively.

A. HYPOTHESIS TESTING

The purpose of hypothesis testing in cytotoxicity studies is to help the investigator in reaching a decision about a population by examining a sample of that population. The hypothesis to be tested or defined, referred to as the ***null hypothesis (H_0)*** or ***hypothesis of no difference***, is usually a statement about one or more populations. In the process of the statistical test, the null hypothesis is either rejected or accepted. If the testing procedure leads to rejection, the data being evaluated are not compatible with the null hypothesis, but are supportive of some other hypothesis, such as that the biochemical changes incurred on the cell are a result of the influence of the chemical. In general, the null hypothesis is established for the explicit purpose of being discredited or supported.

The hypothesis is also concerned with the parameters of the populations in question. The form of the parameters, as data information, determines the choice of tests used for testing. In hypothesis testing of univariate parametric tests, the populations are normally distributed and the data are continuous; that is, each data point has a discrete number and that number has a measurable relationship to other numbers among the populations. In hypothesis testing of nonparametric tests, the data are ranked or categorical. This type of data is not continuous, is presented in the form of contingency tables, and is collected and arranged so that it may be classified according to one or more response categories.

B. PARAMETRIC TESTS

Figures 2 and 3 summarize some of the more frequently used parametric and nonparametric tests in cytotoxicity studies. Parametric testing focuses on estimating or testing a hypothesis about one or more population parameters. A valid assumption of these tests is that the sampled populations should be approximately normally distributed. In addition, there must be a knowledge of the form of the population providing the samples for the basis of the inferences.

1. Regression Analysis

Regression analysis, or linear regression, is one of two statistical tests, along with correlation analysis, that yields information about the relationship between two variables. These variables usually are normally distributed sample populations which take the form of experimental groups. For example, in toxicity testing the experimental groups include the control group, to which no chemical is added, and the treatment groups. Each of the groups forms a data point. The variables in question consist of the measurable response of the cells to a chemical (Y or dependent variable) as a function of the concentration (X or independent variable). Thus regression analysis helps in identifying the probable form of the relationship between these variables, with the ultimate objective of predicting or estimating the value of dependent variable in response to a known value of the independent variable. As in the case of the cytotoxicity experiment, the 50% inhibitory concentration (IC$_{50}$) is estimated based on the regression analysis of the concentration-effect curves. That is, an estimate of the concentration necessary to inhibit 50% of the measured response is then calculated within the confidence intervals. Often, the raw data of concentration-effect relationships are not linear, necessitating transformation of the data into logarithmic value scales. Usually, the *logs* of the concentrations are plotted on the X-axis against the probability (*probit*) scale, such as percentage response of control value (Y-axis). Such manipulations render the information more amenable to linear interpolation and calculation of estimated inhibitory concentrations with greater statistical confidence.

2. Correlation Analysis

Correlation analysis, and the calculation of the correlation coefficient (*r* value), is computed to determine the strength or significance of the regression analysis. Computation of the *r* value gives the relative degree of correlation between the variables at a predetermined level of significance. The largest value that *r* can assume is 1.0, a result when all of the variation in the dependent

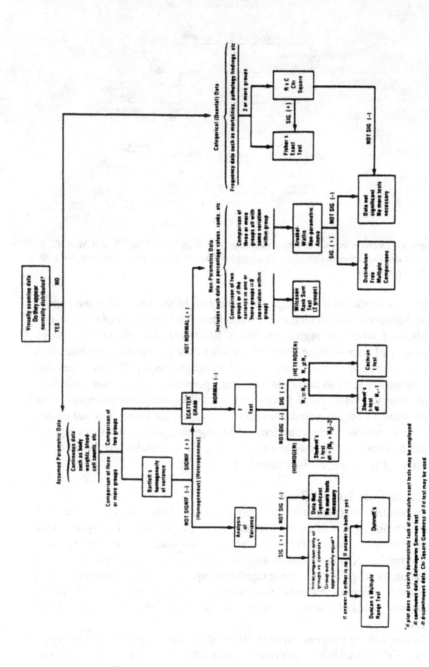

Figure 2. Decision tree for selecting hypothesis-testing procedures. (From Gad, S. and Weil, C. S., *Statistics and Experimental Design for Toxicologists*, 2nd ed., Telford Press, Caldwell, NJ, 1988. 18. With permission.)

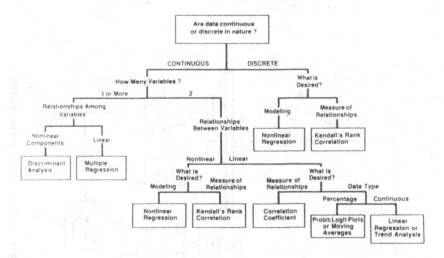

Figure 3. Decision tree for selecting modeling procedures. (From Gad, S. and Weil, C. S., *Statistics and Experimental Design for Toxicologists*, 2nd ed., Telford Press, Caldwell, NJ, 1988, 19. With permission.)

variable is explained by the regression analysis. In this case, all of the experimental observations fall on the regression line. The lower limit of *r* is 0, which is obtained when the regression line is horizontal. The closer the *r* value approaches 1.0 (or −1.0, an inverse correlation), the greater the probability that the variation in the Y values is explained by the regression. This indicates that the regression analysis accounts for a large proportion of the total variability in the observed values for Y. Conversely, a small *r* value (close to zero in either the positive or negative direction) does not support the regression and causes it to be viewed with less confidence. The objective statistical test for significance, however, can be attained by calculating the ***hypothesis test for $\beta = 0$***, the ***analysis of variance***, or the ***student's t-test***.

3. Hypothesis Test for $\beta = 0$

In addition to calculating the correlation coefficient (*r* value) for each concentration-effect curve, the hypothesis test for $\beta = 0$ can be applied. This computation is an alternative test of the null hypothesis of zero slope between two variables, X (log of concentration) and Y (percentage response), and is based on the slope of the sample regression equation. The test statistic, which is normally distributed as with the student's *t*, is calculated and compared to minimum values of t, at $n - 2$ degrees of freedom, at 95 and 99% confidence intervals.

Figure 4 shows the concentration-effect curves that were generated for representative chemicals using [³H]proline incorporation as a measure for cytotoxicity.[1,2] The correlation coefficient, the test statistic (*t* value), and inhibitory concentrations, are calculated from the transform regression analysis of each plot,

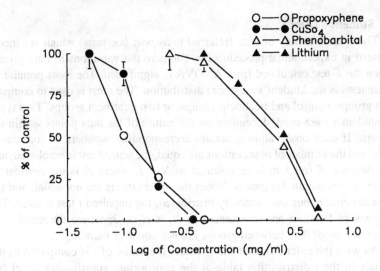

Figure 4. Concentration-effect curves for representative chemicals in cultured rat lung cells using the 24-h [³H]proline assay comparing relative *in vitro* cytotoxicities. Inhibitory concentrations at the 50% level were determined from these plots.

"% of control" vs. "log of concentration". Different sets of dose levels are repeated for each chemical until the test statistic for the plot is significant at the 95% or 99% confidence interval. The test statistic is then compared to minimum values for *t* for the number of pairs of data involved minus two (or *n* – 2 degrees of freedom). If the test statistic is larger than the table value, the relationship between the two groups is significant; i.e., the effect on [³H]proline incorporation is inversely proportional to the change in concentration of the chemical. If the test statistic is smaller than the table value, there is no relationship between the two variables and the true null hypothesis is not rejected.

4. Analysis of Variance (ANOVA) and the *F* test

Analysis of variance compares the total variation present in three or more groups of data, which is partitioned into several components. The data are continuous, independent, and normally distributed. Each component is associated with a specific source of variation, so that each source contributes to a portion of the total variation. The magnitude of the contributions is used to calculate the *F* distribution and compared to a table of *F* values. If the calculated *F* value is greater than the critical value for *F* at a given significance level, then the null hypothesis of equal population variances is rejected. This indicates that the computed *F* value does not represent a rare event brought about by chance, and that there is a significant difference *among* the groups being compared. The Student's *t* test is then computed to determine *which* groups are significantly different. An *F* value which is smaller than the critical value for *F* at a given significance level indicates no significant differences among the groups.

5. Student's *t*-test

There are a variety of tests (referred to as *post hoc* tests) which are incorporated in experimental procedures to compare the variations of data groups when the *F* test calculated from ANOVA is significant. The most popular of these tests is the Student's *t* test, or *t* distribution. The *t* test is used to compare two groups, control and treatment groups, or two treatment groups. Two types of student's *t* are used, depending on the nature of the data points within the groups. If each observation generates corresponding numbers of continuous data, and the number of observations are equal, the *paired* test is employed, and the degrees of freedom is represented as $N - 1$, where N is the number of observations within the groups. When the observations are not equal, and the data are continuous but randomly distributed, the unpaired *t* test is used. The degrees of freedom are represented as $N_1 + N_2 - 2$, where the number of observations of data between groups are not equal or paired.

As with the calculation of the *F* value, the value of *t* is compared to the values in the *t* distribution table at the appropriate significance level (α) and degrees of freedom. If the calculated *t* value is greater than the critical value for *t* at a given significance level, then the null hypothesis of equal sample variances is rejected, indicating that there is a significant difference between the two groups and the changes on the dependent variable are directly or indirectly proportional with changes of the independent variable. A *t* value which is smaller than the critical value for *t* at a given significance level indicates no significant differences between the two groups.

C. NONPARAMETRIC, GOODNESS-OF-FIT TESTS

Nonparametric statistical analysis is exactly analogous to parametric statistical methods, but involves analysis of ranked data. Traditionally known as *goodness-of-fit* tests, these analyses are appropriate when one wishes to decide if an observed distribution of frequencies is compatible with some predetermined or hypothesized distribution. The procedures used for reaching a decision, to reject or accept the null hypothesis, consist of placing the observations into mutually exclusive categories and noting the frequency of observed values in each interval. These frequencies are then compared with available knowledge of normal frequency distributions that would be expected if the samples were obtained from a normal distribution. The discrepancies between what was observed and what is expected are used to calculate the test statistic, which is then compared to minimum values in a table of values for that particular test at a preassigned level of significance.

1. Chi-Square (χ^2)

The chi-square test for independence of classification is a widely used goodness-of-fit test and is based on a cross tabulation of observed frequencies of two (2×2) or more ($R \times C$) variables. This cross-tabulation is also called a contingency table and arranges the discontinuous frequency data in rows (R) of control and treatment groups and in columns (C) of observed and expected frequencies. The degrees of freedom are calculated as $(R-1) \times (C-1)$. Chi-square is applied to normal, binomial, and Poisson distributions, depending on the number of categories. When χ^2 is greater than the tabulated value of χ^2 at the predetermined level of significance (α), the null hypothesis is rejected at that level, and the groups are considered to be significantly different with respect to their treatment protocols. When the null hypothesis is true, the discrepancy between observed and expected frequencies is distributed approximately as the calculated χ^2.

Some underlying assumptions for the chi-square distribution include that the data are univariate and categorical, and that the groups are approximately the same size with at least 50 total observations. As with other statistical analyses, the data should be collected by random, independent sampling methods.

2. Wilcoxon Rank Sum

As with the chi-square, the Wilcoxon rank-sum test is used for comparison of groups of data not normally distributed and whose data falls within certain groups with set ranges of observations. Data for the groups to be compared are initially ranked in increasing order and assigned a rank value. The "sum of ranks" values are calculated and compared to the two limit table values to determine whether the groups are significantly different at a given significance level (α).

3. Kruskal-Wallis Nonparametric ANOVA

As with the parametric analysis of variance, the Kruskal-Wallis non-parametric ANOVA is used to reject or accept the null hypothesis that several group means are equal. It is initially performed on discontinuous, categorical groups of data to determine if any significant differences exist among the groups. The test proceeds by arranging the observations of three or more groups in a single series in order of magnitude from smallest to largest. The observations are then replaced by ranks starting from 1. The ranks assigned to observations in each group are added separately to give the "sum of ranks". The test statistic (H) is computed and compared with a table of H values. When the calculated H value is greater than the table value for the number of observations in the study, the null hypothesis is rejected, suggesting that there is a significant difference between the groups. The test can be followed by applying another nonparametric test, such as the *2 ×2 chi-square* or the *distribution-free multiple comparisons* depending on the number of groups, to determine which groups exhibit significant differences.

The Kruskal-Wallis ANOVA is used for both small and large samples. The test is weakened when many tied ranks occur.

REFERENCES

1. **Barile, F. A., Arjun, S., and Hopkinson, D.**, *In vitro* cytotoxicity testing: biological and statistical significance, *Toxic. In Vitro*, 7, 111, 1993.

2. **Hopkinson, D., Bourne, R., and Barile, F. A.**, *In vitro* cytotoxicity testing: 24- and 72-hour studies with cultured lung cells, *ATLA*, 21, 167, 1993.

3. *Merck Index*, 11th ed., Budavari, S., O'Neil, M. J., Smith, A., and Heckelman, P. E., Eds., Merck and Co., Rahway, NJ, 1989.

4. **Sax, N. I. and Lewis, R. J.**, *Hazardous Chemicals Desk Reference*, 1st ed., Van Nostrand Reinhold, New York, 1987.

5. **Baselt, R. C. and Cravey, R. H.**, *Disposition of Toxic Drugs and Chemicals in Man*, 3rd ed., Year Book Medical Publishers, Chicago, 1989.

6. **Barile, F. A., Arjun, S., and Senechal, J.-J.**, Paraquat alters growth, DNA and protein synthesis in lung epithelial cell and fibroblast cultures, *ATLA*, 20, 251, 1992.

7. **Barnes, Y., Houser, S., and Barile, F. A.**, Temporal effects of ethanol on growth, thymidine uptake, protein and collagen production in human fetal lung fibroblasts, *Toxic. In Vitro*, 4, 1, 1990.

8. **Gad, S. and Weil, C. S.**, *Statistics and Experimental Design for Toxicologists*, 2nd ed., Telford Press, Caldwell, NJ, 1988, 55–59 and 102–104.

9. **Daniel, W. W.**, *Biostatistics: A Foundation for Analysis in the Health Sciences*, 2nd ed., John Wiley & Sons, New York, 1978.

Chapter 11

STANDARDIZATION AND VALIDATION

Bjorn Ekwall and Frank A. Barile

I. INTRODUCTION

A. DEVELOPMENT OF TEST METHODS

In the last decade, a host of *in vitro* test methods have been developed to test various types of nongenetic, general toxicity. These methods are not mechanistically defined, but are empirical measures of all types of toxicity, based on the assumption that their targets are similar to those encountered *in vivo*. Some of these protocols are well-established, such as the use of cell lines to predict local irritancy of bodily implants. Traditionally, these studies have been used to detect the toxicity of plastic or metallic prostheses and dental materials. Other proposed *in vitro* tests have been developed only recently, such as the testing for acute systemic and target organ toxicity (including nervous system, liver, heart, and kidney), local irritancy (skin, eye, and lung), teratogenicity, and various toxicokinetic tests. About 10 to 50 test methods have been developed for each system, totaling 400 to 500 different procedures.[1,2]

Some of these techniques were rapidly adopted by the pharmaceutical and cosmetic industries as preliminary screening tools and used in parallel with conventional animal testing protocols. Based on their comparisons with traditional methods, the newly developed tests are ultimately either accepted or rejected at a later stage in the process. In this way, many methods have been in practical use for prediction of acute toxicity, eye irritancy, phototoxicity, and teratogenicity, based on their ability to screen for potentially toxic substances.[3] In fact, the U.S. Pharmacopoeia recommends cellular tests for local and systemic toxicity of plastic devices as a test of first choice.[4] These methods, however, represent only a small portion of the bulk of procedures recently available and referenced in the literature and have not been subjected to evaluation by industrial concerns. Unlike the procedures which have been evaluated and have withstood the test of time, none of the latter methods have gained wide acceptance by the regulatory, industrial, and scientific communities.

Why are so many *in vitro* methods not accepted into the general toxicity testing scheme? The answer lies in a complex set of scientific and regulatory criteria. In addition, the newer *in vitro* methods have been rigorously, and justifiably, compared to the theoretical basis that has formed the fundamentals of all traditional toxicologic inquiry, with the result often leading to rejection. The rest of this chapter will present an explanation for this reluctance and a proposal for future validation schemes and legislative acceptance.

B. STANDARDIZATION OF TEST METHODS

In general, it is easy for a laboratory to develop and propose an *in vitro* procedure aimed at a specific toxic endpoint. The principle is to analyze a chemical or set of chemicals and their effects on cellular components. These evaluations may involve testing for injury to basal cell or organ-specific functions, alterations to noncellular structures, the inflammatory reactions to the injury, and the toxicokinetics of the chemical in the target tissue. A cell culture system is then constructed to simulate a composite of these factors, or perhaps only one factor, and thus becomes a partial test for the toxicity in question. *In vitro* eye irritancy procedures were developed in this manner as composite protocols, such as the CAM test. Other systems were based on the detection of toxicity on focused targets in the eye, such as tests with corneal epithelium, protein matrix simulating corneal structure, and tests of inflammatory response.[5]

It is difficult and expensive for a single laboratory to prove the validity of a test and establish standards for routine and ubiquitous use. In the case of eye irritancy, the laboratory must prove, to the satisfaction of other laboratories, that the method will screen or predict eye toxicity in humans better than the existing and accepted Draize procedure. (The Draize method is especially noted for its ability to predict toxicity of naturally derived cosmetics.) Essentially, a fledgling procedure will require validation,[3,6,7] i.e., evaluation of the technical reliability of the method as well as the relevance to its stated purpose. A composite method, for instance, will require testing of hundreds of carefully selected chemicals with potential for human eye irritancy. The results would then be compared to those obtained with the Draize method and to existing data for human toxicity. Following this tedious and time-consuming experimentation, the validation process at this point is also cumbersome. If the test has introduced only one component of the target injury, the other components of the *in vivo* data must also be incorporated into the validation scheme. For most laboratories, this labor-intensive endeavor is not a realistic goal, since the availability of funds is often limited. Most researchers, therefore, have performed preliminary and limited evaluation of their methods with the resources available. For instance, 10 to 30 chemicals are tested with the proposed protocol and compared with available animal toxicity data for the same substances — thus, a method is proposed which has undergone a cursory evaluation. The objective, therefore, is not to propose a method to a regulatory agency, but rather to introduce a procedure to the scientific and industrial community, with the hope that it will be further evaluated for practical and general use as part of a toxicity testing protocol.

The bottleneck to implementation of *in vitro* methods into the general toxicity testing schema, therefore, is a lack of well-defined validation proposals, caused primarily by a paucity of resources of individual laboratories devoted to this purpose. Also, human toxicity data for certain types of general toxicity are scarce, which makes it more difficult to accept nonmechanistic *in vitro* tests. Finally, it is important to note that poorly defined validation processes are an impediment to future acceptance of effective *in vitro* testing.

II. SINGLE LABORATORY VALIDATION

A. DEFINITIONS

The term *validation* is derived from the original evaluation of *in vitro* genotoxic tests and has been introduced into the nongenotoxic arena to denote relevance and reliability of a test for a specific purpose.[6,7] *Reliability* refers to the reproducibility of a test within the same laboratory and between laboratories (that is, intra- and interlaboratory variation). *Evaluation* refers to a process whose aim is acceptance of the test by the scientific community, although it does not necessarily imply achievement of this goal. *Validation* suggests that this same goal of acceptance by the scientific community is an absolute, as defined by the regulatory agencies. In spite of this implication, the term is used most often by investigators as a process of evaluating *in vitro* methods, without the demand that the process be complete in order to reach an absolute goal. Several investigators have avoided the term altogether.[8] In this context, the authors suggest that the term refer to a *process* used to evaluate the reliability and relevance of *in vitro* tests, which includes the necessary investigatory steps aimed at acceptance of the method in the general scientific and industrial communities.

B. THE PROCESS OF VALIDATION BY AN INDIVIDUAL LABORATORY

1. Relevance and Reliability

The importance of the relationship between relevance and reliability is necessarily linked to the concept of validation. For instance, a test which demonstrates favorable results for a large number of chemicals implies considerable reliability. In contrast, a very reliable method may not be useful if it has not demonstrated some relevance to the *in vivo* situation. Many recently developed cytotoxicity testing protocols with objective end-point determinations from reputable laboratories are reproducible, and their results do not vary significantly between laboratories. Alternatively, under closer scrutiny, some of these methods have been shown to lack relevance. Therefore, while relevance and reliability are equally important in the validation process, the concept of relevance is the "rate limiting" factor, which allows little tolerance for deviation from the goal of validation. Because of this perception, most single laboratory validation studies have only evaluated relevance, having good reasons to believe that the reliability of the method is sufficient as presented, or could be improved based on the positive evaluation of relevance.

Table 1 lists some examples of single laboratory validation studies performed to date. These pioneering studies do not fulfill all of the criteria discussed above for optimal validation. With the exception of acute systemic toxicity studies,[14-19] most laboratories have used many different compounds

TABLE 1
Examples of Single-Laboratory Validation Studies

Test systems	Type of toxicity	No. of chemicals	Species[a]	Comparison
Cell line toxicity (agar overlay, etc.)	Irritation of muscular implants	1013	Rabbits	Qualitative[25]
Cell line toxicity (KB[b], MIT-24[c])	Acute systemic	200	Rat, mouse	Linear regression[15–18]
MIT-24	Acute systemic	50	Human	Quantitative[5,19]
Cell line toxicity (NRU[d], CB[e])	Eye irritancy	150	Rabbits	Rank correlation[24]
CAM and HET-CAM[f]	Eye irritancy	250	Rabbits	Rank correlation[20,22]
EYTEX™	Eye irritancy	1500	Rabbits	Rank correlation[21,23]

Note: References are noted as superscripts.

[a] Species to which comparisons were performed.
[b] Kenacid blue.
[c] 24-h metabolic inhibition test.
[d] Neutral red uptake.
[e] Coomasie blue.
[f] Hen's egg test-chorioallantoic membrane test.

without regard to grouping chemicals in particular classes. In some studies, however, the numbers represent mixtures of products and materials that are impossible to request for retesting,[20,21] while other reports have not presented all the raw data.[22,23] Most studies used animal data for *in vivo* comparisons. Implantation of materials in rabbits yielded only a semiquantitative response, which explains the qualitative comparison.[20,24] The Draize test scores validated against eye irritancy studies were composites of subjective semiquantitative scores not suitable for validation purposes. Blind testing was performed in only a few of the studies.[17,20,23,24] As with the local implantation scores, the Draize test data only permitted rank correlations, thus negating a necessary *in vitro/in vivo* method of comparison (as defined above). Some of the *in vitro* tests are also scored in a similar manner,[20,22,25] which makes the linear correlation approach difficult. All studies except one[19] compared the *in vitro* test results directly with dose-defined, gross animal toxicity.

Table 1 also presents some examples of validation of methods to test local irritancy of muscular implants, acute systemic toxicity, and eye irritancy, performed by the laboratories developing the procedures. The validation study using cell lines for acute toxicity, and the comparison of the results to rat and mouse LD_{50} data, represents five similar validation approaches, each performed with about 40 chemicals. Each program has reported relatively good correlations with *in vivo* toxicity data (i.e., approximately 70% positive prediction), and this has served to introduce the methods for use as screening tests in a battery of protocols.

How was reliability of tests validated in these studies? Some of the more ambitious laboratories studied reliability with coded substances.[22,23] Others evaluated relevance of methods, which would be more scientifically significant if the results were subjected to multilaboratory methods.[17] Not surprisingly, the remaining studies did not include reliability evaluation, since reliability and relevance have traditionally been considered as separate entities within the same study.

With the exception of tests for local irritancy of muscular implants, single laboratory validation has not gained general acceptance for any of the methods. Nor have they replaced any currently accepted analogous animal protocols. The explanation may reside in the shortsightedness of the validation programs.

Table 2 presents some of the early approaches to the validation process, as well as some of the conditions necessary for refining the procedures. Three important considerations, which may determine the results of single laboratory validation studies, are also discussed below.

2. Selection of Chemicals

The *selection of chemicals* used in validation is critical to the analysis of the results of the study.[9,10] A large number of chemicals should be chosen to represent as many relevant chemical and therapeutic classes as possible. The substances should not be selected based only upon their ease of testing *in vitro* or their presumed cytotoxic action *in vivo*, for the results obtained may risk a technical bias toward the *in vitro* method. The best way to secure a nonbiased selection is to devise a scheme for random selection from a chemical registry (e.g., the NIOSH Registry). A procedure which is validated for a limited number of classes of chemicals will probably not be of much practical value. Another prerequisite for a convincing validation study is to use defined chemicals (that is, by their Chemical Abstracts Service [CAS] formulas or their generic names) rather than proprietary mixtures or coded industrial products. Studies performed with defined chemicals allow for testing and validation to be reproduced in other laboratories, thus acquiring scientific legitimacy. In accordance with the selection of the chemicals and their naming schema, presentation of the data is also important. A study risks losing some credibility if the results of evaluation are published without presentation of a tabulated database. Validation studies without tabulated data from the tests that are performed, run the risk of becoming irrelevant and lack proof of their stated or implied objectives.

TABLE 2
Approaches Aimed at Refining Validation Studies

Early approaches	Recent approaches	Advanced approaches
Few chemicals (10–50)	Larger number of chemicals, various classes	Large number (200), all relevant classes
Easily tested chemicals	Some water-insoluble, volatile chemicals	Randomly selected chemicals
Propriety formulations	Defined chemicals	High purity chemicals
Known in vivo data	Blind *in vivo* data	Coded substances
In vivo data from various protocols/literature, not suitable for comparisons	*In vivo* data from various protocols, suitable for comparisons	Suitable *in vivo* data from the same proto-col and designed experiments
Comparisons to animal toxicity	Comparisons to human toxicity	Parallel comparisons of animal and human toxicity
Qualitative comparisons among *in vitro* tests	Semiquantitative comparisons (rank correlation)	Quantitative linear comparisons
Correlative, empirical comparisons, only	Correlative approach and ad hoc analysis of out-liers	Correlation, plus pre-determined qualitative analysis (mechanisms, toxicokinetics, etc.)
Direct comparison between *in vitro* and gross *in vivo* toxicity	*In vitro* results, plus known data on toxicity or toxicokinetics compared to gross *in vivo* toxicity	*In vitro* results plus supplementary data compared to gross toxicity
One method validated in developing laboratory	Multilaboratory reliability studies of the test	Multilaboratory relevance studies of tests
Centrally operated multi-center validation of a few methods	Such validation of many methods for a certain type of toxicity	Such validation for an unlimited number of methods for many purposes
Multicenter validation to select the best test for a purpose	Multicenter validation to select the best combination of tests (multivariate analysis)	Multicenter validation to select a minimum number of tests to predict various types of toxicity

3. *In Vivo* Data

The ***in vivo data*** used in the program is a second important factor in validation studies. Very few studies have been published concerning the prediction of acute human toxicity by animal lethality tests, or the prediction of human ocular and skin irritancy by the Draize technique (see review by Purchase[9]). It is generally accepted by toxicologists that these correlations are not good (70%), and that the poor comparisons are due to species differences. Ideally, in order to introduce an *in vitro* method for practical use as a procedure for predicting human toxicity, the validation step should use human toxicity data as the sole reference. The problem with human toxicity data is its availability, coupled with the uneven quality of that which is available. Without a standardized database of these values, the *in vitro* methods may be less reliable than the corresponding animal protocols that are used for comparison. If animal results are used as reference, the question of the precision of the animal method will compound the reliability of the cytotoxicity test. The species gap would also be unnecessarily included in the evaluation of the new *in vitro* test. The proper use of animal data is in the comparison to the human data of the same chemicals used in the study, thus establishing a baseline for determining if the *in vitro* method is comparable to the animal results.[3,11–13] Ironically, this may be construed as validation of animal tests, paradoxically forced upon cytotoxicologists.

The bulk of available human data is obtained from select references and poison control centers, and is compiled from clinical case studies, hospital admissions, and emergency room visits. Although this information is not derived according to a systematic method, it represents the most reliable data available for commonly encountered chemicals. Thus, the clinical data should be used as a basis for comparison against *in vitro* values. In addition, for the purpose of statistical evaluation, information obtained for human toxicity is constant and less susceptible to misinterpretation than animal data. Another source of reliable human data is generated through the testing on volunteers of some types of toxicity, such as skin and eye irritancy. Skin tests should produce concentration-effect curves for fixed end-points, while in the case of eye irritancy, testing is limited to a minimal response.

4. Comparison to Other *In Vitro* Methods

The third important procedure in validation is the **method used to compare to *in vitro*** data. Historically, the validation of the mechanistic mutagenicity tests was based on a qualitative *in vitro/in vivo* comparison, primarily because of the difficulty associated with establishing a quantitative relationship. All *in vitro* tests of general toxicity, however, are quantitative and, above all, are thought to simulate quantitative interactions between chemicals and cellular components *in vivo*. The tests therefore should be validated quantitatively by linear regression analysis or similar statistical methods.[3] This analysis must take into consideration the different mechanisms of toxicity and cellular targets associated with individual chemicals and should include categorization of outlines in the design of the validation program. For example, linear regression analysis technically requires exact toxicity data without cut-off results. That is, an IC_{50} value stated to be "greater than x mg/ml" is not meaningful for the analysis.

To date, most validation studies compare the results of *in vitro* toxicity tests directly with some form of human toxicity, such as acute systemic toxicity and skin or eye irritancy. At the same time, most tests were developed to account for only one of many interactions involved in the general toxic effect in humans, including toxicity to targets not measured by the test, and toxicokinetics (absorption, metabolism, distribution, and excretion for systemic toxicity or elimination from the site for local toxicity). When the results from single tests are compared with general toxicity in humans, the comparison is limited to a minor part of the general toxic phenomenon that is being tested, even if this determination is not obscured by toxicokinetics. This type of validation will selectively discard procedures which measure relatively rare effects and, furthermore, cannot discriminate between important supplementary tests and others with similar information. In addition, it is erroneous to compare *in vitro* test results to doses administered in whole animals, whether local or systemic. Instead, comparisons to tissue concentrations can be realized if toxicokinetic data are incorporated in the analysis.[11-13]

5. Comparison to Analogous *In Vivo* Data

It is desirable to compare *in vitro* toxicity test results, which have accompanying toxicokinetic parameters, with *in vivo* toxicity. Historically, earlier studies compared cell toxicity and animal LD_{50}s and found relatively good correlations, indicating comparability of cell toxicity with rodent lethality.[14] In addition, the authors introduced two factors not accounted for by the cytotoxicity tests, i.e., known noncytotoxic lethal action of some compounds and the ability of other compounds to traverse the blood-brain barrier. With this information, the authors demonstrated relevance of the cytotoxicity tests with LD_{50} test results for most compounds. It is conceivable that different *in vitro* toxicity/eye irritancy correlations for different classes of chemicals will depend upon the toxicokinetic information available for that category. Thus it is probable that addition of such information, such as the inclusion of absorption and elimination rates to test batteries, will improve future attempts at correlation.

Recently, the practical and economic advantages of centrally organized multilaboratory validation programs have diminished the interest in performing single laboratory studies.[30,40-42] Both types of validation, however, are based on the same principles outlined above. Future single laboratory validation will always retain its competitive edge with multilaboratory validation programs, primarily due to its comparatively higher flexibility. Thus, well-performed single laboratory validation studies should be as acceptable to regulatory authorities as the corresponding multilaboratory approaches. It is also conceivable that single laboratory validation may also be developed with simple measures, such as literature data, or through the collaboration of two or three laboratories with an interest in the same specialized area.

III. ORGANIZED MULTILABORATORY VALIDATION

Early on, some investigators[5] suggested that several advantages exist with organized collaborative validation efforts, when compared to single laboratory studies. Such multicenter validation was believed to further enhance the ideas launched by the multilaboratory evaluations of the reliability of isolated methods, performed in the 1960s. In its present form, multicenter programs consist of centrally directed testing of the same reference chemicals in many test systems and/or laboratories, with the objective of evaluating both relevance and reliability of methods. There are several types of such studies, which currently are characterized as representing intermediate stages of progress between single laboratory and multilaboratory validation of all possible methods for many types of purposes (see lower half of Table 2). Thus, the development of multicenter policies and procedures is analogous to other important aspects of validation, such as improvement in the selection of chemicals, reference to *in vivo* data, and the methods of correlation.

A. COMPARISON TO SINGLE LABORATORY VALIDATION

Some advantages of multicenter programs over single laboratory validation include:

1. The evaluation of reliability of methods can be performed in parallel with the evaluation of relevance.
2. The credibility of results will increase, compared with single laboratory validation, if an independent group not involved in assay development, has selected the chemicals.
3. Multilaboratory validation is considerably more economical than single laboratory studies because efforts and expenses used to select chemicals and gather *in vivo* data, including the planning and execution of comparisons, can be shared among laboratories.[9]
4. Once computerized procedures for validation of relevance have been established, many protocols can be evaluated, thus limiting the possibility of omitting a valuable approach.
5. Only a multicenter study may directly validate predictivity of combined methods, for instance by computerized multivariate analysis.
6. Only a multilaboratory program can effectively use models of various types of human toxicity (based on knowledge of the most critical targets for toxicity, mechanisms of action, and key toxicokinetic data) and thus determine if diverse incoming data can contribute to parts of these models.
7. The program can attempt to model various *in vitro* data as they are received, according to their application to different types of toxicity, such as acute systemic, eye, and skin irritancy, and teratogenicity. As a result, some of the *in vitro* results may be predictive in distinct models. Thus, a minimum number of different *in vitro* tests able to predict various types of human toxicity could be defined. Such validation can be performed neither by single laboratories, nor by multilaboratory studies focusing on one type of toxicity.

8.   The results of different laboratories, using the same set of reference chemicals in different *in vitro* toxicity tests, can be compared. Thus the advantage of a multilaboratory program is the interlaboratory comparison of methods using different targets and toxicity criteria. For instance, differential cytotoxicity studies, as performed by individual investigators,[27,28] may be recognized on a large scale. Such comparisons may not benefit the purposes of *in vitro/in vivo* validation, but will probably contribute significantly to our understanding and development of *in vitro* toxicology.

B. TYPES OF VALIDATION PROCESSES

In recent years, several organizations have instituted multilaboratory validation studies. This is a result of several factors, including (1) recognition of validation as a bottleneck to the practical introduction of ethical, economical, and target-related *in vitro* methods in toxicity testing; (2) recognition of the advantages of multilaboratory validation; and (3) societal and political pressures to introduce alternatives to specific animal tests, such as the Draize eye irritancy test.

Table 3 presents ongoing multilaboratory validation studies, which have been briefly described in a recent review.[3] As one would expect from newly initiated, centrally organized, second generation validation studies, the methods to select chemicals, to collect *in vivo* data, and the correlation strategies, are generally more refined. With the exception of the ZEBET program,[29] most of the approaches are intended to be pilot studies, as indicated by the small number of chemicals included at this stage. Generally the chemical classes tested have been selected with greater consideration for adequate representation of different classes of chemicals, and the tests are performed with coded substances. The organizers have carefully defined the animal data and, in some cases, the information has been newly generated for this purpose.[8] Human data are used in several studies. In contrast to the general eye irritancy programs, the correlative approach is limited by the Draize test data consisting primarily of rank correlations. The CFTA study uses a refined "concordance analysis" and linear regression.[8] The FRAME and MEIC studies use linear regression, supplemented by predetermined analysis of outliers and simple toxicokinetic modeling.[10,30] Two studies have focused their attention on reliability evaluation and interlaboratory comparisons of *in vitro* data, rather than on evaluation against *in vivo* data for relevance.[26,31] Of the 150 FRAME chemicals selected, only 56 were evaluated for relevance against rodent LD_{50}s.[17]

An examination of the number of methods outlined in Table 3 reveals that there are two types of multilaboratory validation studies — validation based on the reliability of a few methods and the multilaboratory/multimethod programs.

1. Validation Based on the Reliability of a Few Methods

The first type of study focuses on the evaluation of reliability of a few (often two) methods, which are then used in several laboratories to test the reference

TABLE 3
Current Multilaboratory Validation Programs

Organization		Type of toxicity[a]	No. of chemicals (labs[b])	No. of methods	Species[c]
PMA	U.S. Pharmaceuticals Manufacturers Assoc.	Local IM		1 (10)[27]	
FRAME	U.K. Fund for Relacement of Animals in Medical Experiments	AS Eye	150 (56)	6	Mouse[14,26,28-31]
MEIC	Scandinavian Society of Cell Toxicology	AS Skin	50	200	Man[5,25]
MRES	French Ministry of Research and Education	Liver	30	2 (6)	Rodent, man[32]
SDA	U.S. Soap and Detergent Association	Eye	23	14	Rabbit[33]
CFTA	U.S. Cosmetics, Fragrances and Toiletries Association	Eye	10 (10)	23	Rabbit[9]
EC	Commission of European Communities	Eye	21	5	Rabbit[34]
ZEBET	German Ministry of Health humans	Eye	35 (200)	2 (35)	Rabbit[35], humans
OPAL	French organization for Assistance to Laboratory Animals	Eye	40[d]	2	Rabbit[36]
JSAAE	Japanese Society for Alternatives to Animal Experiments	Eye	52	2	Rabbit[37]

Note: References are noted as superscripts.

[a]   IM = intramuscular, AS = acute systemic, eye = eye irritancy, skin = skin irritancy.
[b]   Number of laboratories used to evaluate interlaboratory reproducibility.
[c]   Species of comparison.
[d]   Completed 1991.

chemicals.[26,29-34] To a limited extent, these inquiries have also evaluated relevance of tests with *in vitro/in vivo* comparisons. This examination depends on the variation in the quality of *in vivo* data and the total number of chemicals used in the reliability study. In general, these reliability-focused, ***multilaboratory studies of a few methods*** represent a relatively primitive approach for the determination of relevance. In contrast to single laboratory validation, these programs only analyze techniques as single and ultimate predictors of *in vivo* toxicity. The capacity of a protocol to predict human toxicity, as measured in one study, will be difficult to compare with an analysis of other tests, as measured in other similar studies. In the past, single laboratory validation programs have been criticized for promoting acceptance of methods prematurely, before other potentially valuable techniques had been presented. The same criticism now applies for multilaboratory validation of a few methods. Since the relevance of a procedure is not associated with or predicted by the reliability, the fact that these studies have focused on reliability evaluation is not rational nor economic. Thus, resources are expended to screen for technically reliable tests, which at a later stage may prove to have poor relevance, when competing with other trials.

2. Multilaboratory, Multimethod Programs

In contrast to validation based on a few methods, Table 3 illustrates some of the validation efforts using a larger number of methods, as well as other similar characteristics.[8,10,34,35] Although these studies do not neglect the reliability evaluation, they are primarily concerned with evaluation of relevance of many methods at the same time, and may be called *relevance-focused multilaboratory studies of multiple methods*. These programs always introduce an element of competition among the methods, while also varying the multiple method approaches. The setup for such studies could be wholly dependent on the cooperation of other laboratories that develop or use the test. Such programs are designed as control centers for the listing of chemicals to be tested by the laboratories and the coordination of the data,[10] or they are involved with the distribution of coded chemicals to participating laboratories.[35] Alternatively, these investigations can be done in contract laboratories, without the need for developing new facilities. The studies are open to investigators who have developed the protocols and have not been solicited to participate in the project. Some programs appear to have traits of both reliability focused programs and multilaboratory validation.[31,35]

Only the multiple method multilaboratory programs, in sharp contrast to both single laboratory and reliability focused multilaboratory programs, will benefit from the advantages detailed above (Section III.A). The multiple method studies also incorporate most of the advanced aspects of validation (as listed in Table 2).

The CFTA and MEIC programs are typical multiple method projects which have sought to improve upon the multimethod validation strategies. Some of their basic principles of organization and approaches to validation are described.

a. The CFTA Program

The Cosmetics, Fragrances, and Toiletries Association (CFTA) program is based upon the preliminary evaluation of 23 methods for the prediction of local eye irritancy. Evaluation is accomplished by comparison with Draize test data. Different classes of chemicals and formulations of interest to the cosmetic industry are used in a stepwise evaluation. The first phase analyzes a series of hydroalcoholic formulations, followed by a second phase which analyzes a series of oil-water emulsions. The testing continues in this fashion. Methods which are not completely successful at predicting the first phase of the study for the list of substances are not necessarily excluded from subsequent testing.[8]

b. The MEIC Program

In 1987, the Scandinavian Society of Cell Toxicology invited international laboratories to participate in the *Multicenter Evaluation for In Vitro Cytotoxicity* (MEIC). The program aims at an evaluation of the predictive power of various tissue culture tests of acute cytotoxicity for human acute systemic toxicity. All participants are requested to test the cytotoxicity of a list of 50 reference chemicals by the various methods used in their laboratories. The chemicals have been selected on the basis of available data on human toxicity, such as lethal dosage, toxic blood concentrations, effects on target organs, and toxicokinetics. Conventional animal toxicity data are also available for the chosen chemicals. The IC_{50} values generated in the participating laboratories, and requested by and submitted annually to the MEIC committee, are analyzed, validated, and published before the next yearly deadline. The results from the tests are then compared with the precise data on acute human and animal toxicity and with other laboratories.

Currently the program is developing studies of local irritancy involving human volunteers. All *in vitro* methods are subject to evaluation, especially those which show promise of relevance to various types of human general toxicity, as judged by participating laboratories. Furthermore, all results generated by participating, volunteer laboratories are evaluated for their combined relevance to all types of toxicity, including acute systemic, chronic systemic, local skin, and various target organ toxicities. The methods are judged according to their ability to demonstrate optimum correlation with corresponding human toxicity data. A cell test is evaluated, however, not only on its correlation with human and animal toxicity, but also on its ability to combine with other cellular methods from different laboratories. It is anticipated that these comparisons will ultimately result in the selection of a battery of tests which give the best prediction of human toxicity. The evaluation process used in the MEIC study is conducted in three consecutive phases, corresponding to the block of chemicals being tested (Table 4). That is, phase I is designed to evaluate chemical numbers 1 through 10; phase II, chemical numbers 11 through 30; and phase III, numbers 31 through 50.[3,10]

C. A COMPARISON OF THE METHODOLOGIES

The CFTA study validates specific tests for certain purposes, i.e., tests developed to predict eye irritancy. The nature of this multilaboratory program allows it to directly evaluate the best combination of tests with optimum reliability, although this does not necessarily imply the best combination of well-correlated methods. It also evaluates tests according to their inclination for partial prediction, along with other partial toxicity and toxicokinetic data and tests which fit toxicokinetic modeling of eye irritancy. Since it has focused on only one type of toxicity, the investigation may not incorporate *in vitro* tests developed for other purposes, such as for acute systemic, skin, and teratogenic toxicity. These tests were not initially in the program, due to presumed irrelevance. Thus, the study cannot contribute to

TABLE 4
First 50 Reference Chemicals of the
MEIC Project

1. Acetaminophen	26. Arsenic trioxide
2. Aspirin	27. Cupric sulfate
3. Ferrous sulfate	28. Mercuric chloride
4. Diazepam	29. Thioridazine HCl
5. Amitriptyline	30. Thallium sulfate
6. Digoxin	31. Warfarin
7. Ethylene glycol	32. Lindane
8. Methyl alcohol	33. Chloroform
9. Ethyl alcohol	34. Carbon tetrachloride
10. Isopropyl alcohol	35. Isoniazid
11. 1,1,1-Trichloroethane	36. Dichloromethane
12. Phenol	37. Barium nitrate
13. Sodium chloride	38. Hexachlorophene
14. Sodium fluoride	39. Pentachlorophenol
15. Malathion	40. Verapamil HCl
16. 2,4-Dichlorophenoxyacetic acid	41. Chloroquine phosphate
17. Xylene	42. Orphenadrine HCl
18. Nicotine	43. Quinidine sulfate
19. Potassium cyanide	44. Diphenylhydantoin
20. Lithium sulfate	45. Chloramphenicol
21. Theophylline	46. Sodium oxalate
22. Dextropropoxyphene HCl	47. Amphetamine sulfate
23. Propranolol HCl	48. Caffeine
24. Phenobarbital	49. Atropine sulfate
25. Paraquat	50. Potassium chloride

future *in vitro* toxicity testing which employs a minimal number of tests for modeling various types of human toxicity and which may have application for local irritancy. In contrast, the MEIC study is able to investigate the pertinence of this prospective approach by the use of a variety of *in vitro* tests to model different types of toxicity. This foresight of the MEIC study is very practical since many *in vitro* tests developed today have the ability to model most types of general toxicity, together with tests of toxicity to specific targets. Such tests include basal cytotoxicity tests, methods using protein denaturing as indicators of toxicity, and tests on metabolism and distribution of chemicals, such as epithelial passage. The speculation assumes that modeling of diverse types of toxicity is more dependent on the relative tuning of similar sets of tests, than radically different tests within the sets.

It is very important that validation in the future is not hampered by theoretical considerations of the best way to achieve this goal. The tabulation of steps of refinement, as outlined in Table 2, is presented as a set of generally applicable rules, nonetheless attainable. The most effective future validation will probably require studies directed at all levels, including single laboratory, reliability-oriented multilaboratory, and multiple-method multilaboratory perspectives focused on one specific toxicity, as well as multiple-method multilaboratory programs focusing on a broader scope. Combinations of different objectives could be evaluated in different studies.[3] For instance, several methods could be analyzed for reliability in some programs, and at the same time be evaluated for specific or general relevance in other programs. It is important to note that it may not be possible to combine some objectives in one study, such as selection of chemicals, choice of *in vivo* data, and demands on correlative methods (Table 2). For instance, human data on systemic toxicity may not be compatible with a particular selection of some chemicals. There may necessarily be some conflict of goals in the selection of optimum objectives in the scheme of validation proposals.

IV. PROPOSED VALIDATION PROGRAMS

A. INTRODUCTION

To date, many *in vitro* toxicity tests have been proposed and have accumulated without the benefit of arriving at general acceptance. This situation is due in part to the difficult tasks and characteristics associated with single laboratory validation. The organization of multilaboratory validation programs has been prompted by *in vitro* toxicologists as a possible remedy for the problem. In addition, animal activists and industrial and governmental agencies, including the bureaus of OECD and EC, are engaged in the promotion of the development of *in vitro* toxicity tests by organized, well planned studies. Some organizations, such as FRAME (Fund for Replacement of Animals in Medical Experiments, U.K.), CAAT (Center for Alternatives to Animal Testing, Baltimore) and ERGATT (European Research Group for Alternatives to Animal Testing)(Table 3) have coordinated their efforts into proposals for future multicenter validation. The first proposed multicenter programs were outlined by FRAME,[26,35] which was soon followed by special conferences on validation arranged by the CAAT group.[6,36,37] Later, the CAAT and ERGATT groups, in collaboration with FRAME, arranged a consensus conference on validation, the result of which was the development of a set of guidelines for validation of toxicity tests, presented as "scientific validation".[7] The CAAT/ERGATT conference was followed by the ERGATT/CEC (Commission of the European Communities) conference on procedures for acceptance of validated tests.[38] Thus, the CAAT/ERGATT report represents the most recent proposal for a relevance focused multimethod approach to validation.

B. THE CAAT/ERGAAT DOCUMENT
1. Overview

The CAAT/ERGATT document has been promoted as an authoritative report setting the guidelines for validating any type of *in vitro* test, including both genetic and nongenetic methods.[7]

The scientific validation process was outlined at the workshop and consists of four stages:

1. Intralaboratory assessment to evaluate reproducibility and preliminary relevance with the use of 5 to 10 compounds, in the laboratory of origin
2. Interlaboratory appraisal to evaluate variation and reproducibility, with the use of 5 to 10 and 10 to 20 substances, in at least four laboratories
3. Procedures established to test the development of the database with respect to relevance, involving the use of 200 to 250 chemicals tested in a small number of laboratories
4. Evaluation of performance of single tests and their applicability as part of test batteries

All substances are coded and the methods are selected according to ability for solving stated problems. In addition, procedures are selected based on their need in relation to the availability of other methods. Any procedure may be rejected from the process if it fails intralaboratory assessment (less than 70% reproducibility) and interlaboratory assessment (less than 70 to 80% interlaboratory reproducibility).

Furthermore, according to the CAAT/ERGAAT document, "scientific validation" studies should have descriptors, including:

1. Level of toxicological assessment
2. The required testing activity

According to the levels of toxicological assessment, four levels have been established as satisfying the minimum criteria for an independent method: (A) toxic potential; (B) toxic potency; (C) chemical hazard classification; and, (D) risk to specific populations. The required testing activity classifies the procedure according to one of three levels: (a) as a screening; (b) as an adjunct; or, (c) as a replacement test.

These descriptors apply to organized studies as well as individual methods. They were formulated in order to influence the stringency of the criteria of test performance, and to allow the inclusion of valuable methods which may not satisfy all of the criteria. Thus, screening of potential toxicity may be accomplished by including less relevant tests than, for example, the replacement of a particular animal procedure traditionally used for risk evaluation. A replacement test which evaluates the risk to a specific population (for example, level D/c above) is as valuable as a screening test for toxic potential (level A/a).

2. Vital Components of a Validation Program

The Document discussed three vital parts of any validation study as described above (Section II.B); that is, the *selection of chemicals,* the *choice of in vivo reference data,* and the *methods of comparisons*. Selection of chemicals was not described in detail. The recommendation defined that structures of chemicals should be presented as part of the study, so that validated tests will not be presumed to encompass chemical structures not included in the validation. This method lacks sophistication. Not selecting chemicals according to their *in vivo* data will force a lack of applicability and acceptance of some tests as well as a very limited usefulness of validated tests. Thus, such trials will lack the precision of predicting acute human and animal toxicity, when compared to corresponding conventional animal procedures.

For reference data, the use of human data was acknowledged and was considered to be fully compatible with the proper utilization of existing animal data.

Concerning methods for comparison of reliability, qualitative, stochastic methods were discussed, without defining methods capable of evaluating dose-response relationships. Similarly, the methods to evaluate relevance were based on binary classification schemes, without addressing the problems of continuous responses. Scrutiny of close mechanistic relationships between *in vitro* tests and *in vivo* toxicity was recommended, but methods to perform such comparisons were not defined. The selection of batteries of methods was based on statistical approaches, such as Bayesian methods of validated isolated tests, as well as incorporation of multivariate regression analysis. Modeling of *in vivo* toxicity by combining cytotoxicity and toxicokinetic data or other modeling with combined *in vitro* data[30] were not included in the proposed process.

A unique feature of the Document stated that the processes described would be an absolute measure of *certification* of test validity, especially attainable by adhering to the descriptors. This consensus implies that the validation process is not open to competition but is designed to generate certified tests.

The "scientific validation" scheme appears to mimic the reliability focused multicenter validation studies, similar to the programs described in Table 3. As such, it may reject potentially relevant tests at an early stage, due to low reliability performance of the test during its development. Also, reliable but potentially irrelevant and redundant methods will be included as the data base is established. There is also the risk of preventing plausibly relevant methods from entering the process, as a result of frivolous prejudgment of controversial hypotheses.

Thus the thought process used to construct a relatively typical validation program does not have a firm scientific foundation. The program suggests a special, authoritative status for the CAAT/ERGATT proposal over other validation efforts, while in fact, some current projects, such as the relevance focused multicenter programs, may be more scientifically advanced. Equally disputable is the claim that the CAAT/ERGATT proposal is able to certify tests as validated and ready for regulatory acceptance. Such claims cannot be made for the results of any validation program, if the study is to be seriously considered by the regulatory and administrative agencies.

More recently, Goldberg et al.,[43] through the Validation and Technology Transfer Committee of the Johns Hopkins Center for Alternatives to Animal

Testing Advisory Board, have proposed a framework for fostering the validation of new methods. This framework is necessary for the effective dissemination of information from recent developments in the laboratory to the research community, and includes four essential validation resources: chemical banks, cell and tissue banks, a data bank, and reference laboratories. In addition, the proposal is divided into three major units: test development, validation, and acceptance. Among the features which are identified as important components of the validation unit are: validation resources, such as chemical, data cell and tissue banks, and reference laboratories; scientific advisory board review panels, independent validation programs, and publication in the peer-reviewed literature. In fact, the Committee suggests that only when a test is accepted in a major scientific peer-reviewed journal can the method be considered truly validated.[43]

It is important to note that the progress of *in vitro* cytotoxicity testing relies on the continuous development of better strategies to evaluate new tests, rather than as a set of inflexible guidelines for the inclusion or exclusion of protocols. Thus, planned chemical databanks for validation purposes should be administered by independent scientific bodies, rather than any particular organization. These groups must be free from vested interests, with their services directed toward the development of scientifically valid programs.

REFERENCES

1. **Zbinden, G.,** Reduction and replacement of laboratory animals in toxicological testing and research, Interim Report 1984–1987, *Biomed. Environ. Sci.*, 1, 90, 1988.
2. **Zbinden, G.,** Acute toxicity testing: public responsibility and scientific challenges, *Cell Biol. Toxicol.*, 2, 325, 1986.
3. **Ekwall, B., Bondesson, I., Hellberg, S., Hogberg, J., Romert, L., Stenberg, K., and Walum, E.,** Validation of in vitro cytotoxicity tests — past and present strategies, *ATLA*, 19, 226, 1991.
4. **U.S. Pharmacopoeia,** 22nd rev., U.S Pharmacopoeial Convention, Inc., Mack Publishing Co., Easton, PA, 1990, 1572.
5. **Ekwall, B.,** Screening of toxic compounds in mammalian cell cultures, *Ann. N.Y. Acad. Sci.*, 407, 64, 1983.
6. **Frazier, J. M.,** Scientific criteria for validation of *in vitro* toxicity tests, Paris: Organization of Economic Cooperation and Development (OECD), environmental monograph no. 36, 1990.
7. **Balls, M., Blaauboer, B., Brusick, D., Frazier, J., Lamb, D., Pemberton, M., Reinhardt, C., Roberfroid, M., Rosenkranz, H., Schmid, B., Spielman, H., Stammati, A., and Walum, E.,** Report and recommendations of the CAAT/ERGATT workshop on the validation of toxicity test procedures, *ATLA*, 18, 313, 1990.
8. **Gettings, S. D., Dipasquale, L. C., Bagely, D. M., Chudkowski, M., Demetrulias, J. L., Feder, P. I., Hintze, K. L., Marenus, K. D., Pape, W., Roddy, M., Schnetzinger, R., Silber, P., Teal, J. J., and Weise, S. L.,** The CFTA evaluation of alternatives program: an evaluation of *in vitro* alternatives to the Draize primary eye irritation test, (Phase I) Hydroalcoholic formulations; a preliminary communication, *In Vitro Toxicol.*, 3, 293, 1990.

9. **Purchase, I.**, An international reference chemical data bank would accelerate the development, validation, and regulatory acceptance of alternative toxicology tests, *ATLA*, 18, 345, 1990.

10. **Bondsesson, I., Ekwall, B., Hellberg, S., Romert, L., Stenberg, K., and Walum, E.**, MEIC — a new international multicenter project to evaluate the relevance to human toxicity of *in vitro* cytotoxicity tests, *Cell Biol. Toxicol.*, 5, 331, 1989.

11. **Barile, F. A., Arjun, S., and Hopkinson, D.**, *In vitro* cytotoxicity testing: biological and statistical significance, *Toxicol. in Vitro*, 7, 111, 1993.

12. **Hopkinson, D., Bourne, R., and Barile, F. A.**, *In vitro* cytotoxicity testing: 24- and 72-hour studies with cultured lung cells, *ATLA*, 21, 167, 1993.

13. **Hellberg, S., Eriksson, L., Jonsson, J., Lindgren, F., Sjostrom, M., Wold, S., Ekwall, B., Gomez-Lechon, M. J., Clothier, R. H., Triglia, D., Barile, F. A., Nordin, M., Tyson, C. A., Dierickx, P., Shrivastava, R., Tingsleff-Skaanild, M., Garza-Ocanas, L., and Fiskesjo, G.**, Analogy models for prediction of human toxicity, *ATLA*, 18, 103, 1990.

14. **Ekwall, B.**, Correlation between cytotoxicity *in vitro* and LD_{50} values, *Acta Pharmacol. Toxicol.*, 52, 80, 1983.

15. **Walum, E. and Peterson, A.**, On the application of cultured neuroblastoma cells in chemical toxicity screening, *J. Toxicol. Environ. Health*, 13, 511, 1984.

16. **Halle, W. and Gores, E.**, Vorhersage von LD_{50}-werten mit der Zellkultur [Quantitative prediction of LD_{50} with use of cell cultures] (German), *Pharmazie*, 42, 245, 1987.

17. **Clothier, R. H., Hulme, L. M., Smith, M., and Balls, M.**, Comparison of the *in vitro* cytotoxicities and acute *in vivo* toxicities of 59 chemicals, *Mol. Toxicol.*, 1, 571, 1989.

18. **Fry, J. R., Garle, M. J., Hammond, A. H., and Hatfield, A.**, Correlation of acute lethal potency with *in vitro* cytotoxicity, *Toxicol. in Vitro*, 4, 175, 1990.

19. **Ekwall, B.**, Preliminary studies on the validity of *in vitro* measurement of drug toxicity using HeLa cells. III. Lethal action of man of 43 drugs related to the HeLa cell toxicity of the lethal drug concentrations, *Toxicol. Lett.*, 5, 319, 1980.

20. **Bagley, D. M., Rizvi, P. Y., Kong, B. M., and DeSalva, S. J.**, An improved CAM assay for predicting ocular irritation potential, in *Alternative Methods in Toxicology*, Vol. 6, Goldberg, A. M., Ed., Mary Ann Liebert, New York, 1988, 131.

21. **Soto, R. J., Servi, M. J., and Gordon, V. C.**, Evaluation of an alternative method for ocular irritation in *Alternative Methods in Toxicology*, Vol. 7, *In vitro* toxicology: new directions, Goldberg, A. M., Ed., Mary Ann Liebert, New York, 1989.

22. **Luepke, N-P. and Wallat, S.**, HET/CAM: reproducibility studies, in *Alternative Methods in Toxicology*, Vol. 5, Goldberg, A. M., Ed., Mary Ann Liebert, New York, 1987, 353.

23. **Gordon, V. C., Kelly, C. P., and Bergman, H. C.**, Application of the EYTEX™ method, *Toxicol. In Vitro*, 4, 314, 1990.

24. **Shopsis, C., Borenfreund, E., and Stark, D. M.**, Validation studies on a battery of potential *in vitro* alternatives to the Draize test, in *Alternative Methods in Toxicology*, Vol. 5, Goldberg, A. M., Ed., Mary Ann Liebert, New York, 1987, 31.

25. **Autian, J.**, The new field of plastics toxicology: methods and results, in *Crit. Rev. Toxicol.*, 2, 1, 1973.

26. **Balls, M. and Bridges, J. W.**, The FRAME research program on *in vitro* cytotoxicology, in *Alternative Methods in Toxicology*, Vol. 4, Goldberg, A. M., Ed., Mary Ann Liebert, New York, 1984, 62.

27. **Ekwall, B. and Acosta, D.**, *In vitro* comparative toxicity of selected drugs and chemicals in HeLa cells, Chang liver cells, and rat hepatocytes, *Drug Chem. Toxicol.*, 5, 229, 1987.

28. **Ekwall, B. and Ekwall, K.**, Comments on the use of diverse cell systems in toxicity testing, *ATLA*, 15, 193, 1988.

29. **Kalweit, S., Besoke, R., Gerner, I., and Spielmann, H.**, A national validation project of alternatives to the Draize rabbit eye test, *Toxicology In Vitro*, 4, 702, 1990.

30. Ekwall, B., Bondesson, I., Castell, J. V., Gomez-Lechon, M. J., Hellberg, S., Hogberg, J., Jover, R., Ponsoda, X., Romert, L., Stenberg, K., and Walum, E., Cytotoxicity evaluation for the first ten MEIC chemcials: acute lethal toxicity in man predicted by cytotoxicity in five cellular assays and by oral LD_{50} in rodents, *ATLA*, 17, 83, 1989.

31. Clothier, R. H. and Balls, M., Validation of alternative toxicity tests: principles, practices, and cases, *Toxicol. In Vitro*, 4, 692, 1990.

32. Fautrel, A., Chesne, C., Guillouzo, A., de Sousa, G., Placidi, M., Rahmani, R., Braut, F., Pichon, J., Hoellinger, H., Vintezou, P., Diarte, I., Melcion, C., Cordier, A., Lorenzon, G., Benicourt, M., Vannier, B., Fournex, R., Peloux, A. F., Bichet, N., Gouy, D., Cono, J. P., and Lounes, R., A multicenter study of acute *in vitro* cytotoxicity in rat liver cells, *Toxicology In Vitro*, 5, 543, 1991.

33. Blein, O., Adolphe, M., Lakhdar, B., Cambar, J., Gubanski, G., Castelli, D., Contie, C., Hubert, F., Latrille, F., Masson, P., Clouzeau, J., Le Bigot, J. F., De Silva, O., and Dossou, K. G., Correlation and validation of alternative methods to the Draize eye irritation test (OPAL project), *Toxicology In Vitro*, 5, 555, 1991.

34. Booman, K. A., DeProspo, J., Demetrulias, J., Driedger, A., Griffith, J. F., Grochoski, G., Kong, B., McCormic, W. C. III, North-Root, H., Rozen, M. G., and Sedlak, R. I., The SDA alternatives programme: comparison of *in vitro* data with Draize test data, *J. Toxicol. Cutan. Ocular Toxicol.*, 8, 35, 1989.

35. Balls, M., Riddell, R. J., Horner, S. A., and Clothier, R. H., The FRAME approach to the development, validation, and evaluation of *in vitro* alternative methods, in *Alternative Methods in Toxicology*, Vol. 5, Goldberg, A. M., Ed., Mary Ann Liebert, New York, 1987, 45.

36. Frazier, J. M., Gad, S. C., Goldberg, A. M., and McCulley, J. P., Validation, in *Alternative Methods in Toxicology*, Vol. 4, Goldberg, A. M., Ed., Mary Ann Liebert, New York, 1986, 113.

37. Frazier, J. M., The validation approach of the Johns Hopkins Center for Alternatives to Animal Testing, in *Alternative Methods in Toxicology*, Vol. 5, Goldberg, A. M., Ed., Mary Ann Liebert, New York, 1987, 87.

38. Balls, M., Botham, P., Cordier, A., Fumero, S., Kayser, D., Koeter, H., Koundakjian, P., Lindquist, N.-G., Meyer, O., Pioda, L., Reinhardt, C., Rozemond, H., Smyrniotis, T., Spielman, H., Van Looy, H., Van der Venne, M.-T., and Walum, E., Report and recommendations of an international workshop on promotion of the regulatory acceptance of validated non-animal toxicity test procedures, *ATLA*, 18, 339, 1990.

39. Gad, S. C., Acute ocular irritation evaluation: *in vivo* and *in vitro* alternatives and making them the standard fortesting, in *Benchmarks: Alternative Methods in Toxicology*, Mehlman, M. A., Ed., Princeton University Press, Princeton, NJ, 1989, 137.

40. Wallace, K. A., Harbell, J. H., Accomando, N., Triana, A., Valone, S., and Curren, R. D., Evaluation of the human epidermal keratinocyte neutral red release and neutral red uptake assay using the first 10 MEIC test materials, *Toxicol. In Vitro*, 6, 367, 1992.

41. Dierickx, P. J. and Ekwall, B., Long term cytotoxicity testing of the first 20 MEIC chemicals by the determination of the protein content in human embryonic lung cells, *ATLA*, 20, 285, 1992.

42. Ekwall, B., Barile, F. A., Clothier, R. H., Gomez-Lechon, M.J., Hellberg, S., Nordin, M., Stadtlander, K., Triglia, D., Tyson, C. A., and Walum, E., Prediction of acute human lethal dosage and blood concentrations of the first 10 MEIC chemicals by altogether 25 cellular assays, *In Vitro Cell. Dev. Biol.*, 26, 26A, 1990.

43. Goldberg, A. M., Frazier, J. M., Brusick, D., Dickens, M. S., Flint, O., Gettings, S. D., Hill, R. N., Lipnick, R. L., Renskers, K. J., Bradlaw, J. A., Scala, R. A., Veronisi, B., Green, S., Wilcox, N. L., and Curren, R. D., Framework for validation and implementation of *in vitro* toxicity tests, *In Vitro Cell. Develop. Biol.*, 29A, 688, 1993.

Chapter 12

CELL CULTURE OR ANIMAL TOXICITY TESTS? OR BOTH?

I. ANIMAL EXPERIMENTATION AND ANIMAL RIGHTS

A. ANIMAL REGULATIONS

Outside the building where a recent stockholders meeting of the U.S. Surgical Corporation was being conducted in New York City (June 1990), animal rights activists from the Friends of Animals organization protested the use of surgical staples on dogs (WCBS News Radio, New York). The company manufactures the devices and trains surgeons and salesmen on their use. Without this training on dogs, surgeons would not be capable of performing procedures in situations which do not lend themselves to practice or experiment on humans, such as in emergency rooms. These and similar episodes by animal rights activists have brought animal experimentation to the forefront of science news. Their efforts have forced the science research and education communities to acknowledge responsibility for, as well as to justify, the use of animals in scientific laboratories, and especially in toxicity testing. While trying to comply with the regulations established by the *U.S. Animal Welfare Act* and its amendments (1985), some research laboratories have been forced to terminate their efforts because of the high cost of maintaining and providing for animals as determined by the Act. In response, the U.S. Public Health Service, for example, has established the Office of Animal Research and the Office of Science Education, which promote science education, coordinate policies, and distribute information on the use of animals in research. In addition, some academic and pharmaceutical laboratories have aggressively developed alternative methods to animal experimentation. Historically, the use of animals in biomedical research has provided, and will continue to provide, valuable information about chemicals, drugs, and products used in society. Consequently, the burden of developing suitable protocols to test the safety and efficacy of thousands of drugs and chemicals which are used in medicine or introduced by the chemical industry is incumbent upon the scientific community.

B. RELATIONSHIP OF *IN VITRO* TESTS TO ANIMAL EXPERIMENTS

No single method can be expected to cover the complexity of general toxicity in man. The response of the mammalian organism to a chemical is known to involve various physiologic targets and a variety of complex toxicokinetic factors. The chemicals can induce injury through an assortment of toxic mechanisms. In addition, there are questions which have emerged as a result of previous, preliminary evaluation studies of the predictivity of general toxicity by *in vitro* tests:

1. Can *in vitro* tests be used to model different types of quantitative general toxicity, such as acute systemic toxicity or local irritancy? This testing may require several relevant systems, such as human hepatocytes, heart, kidney, lung, and nerve cells, and other cell lines of vital importance. Can the information obtained from these cells (i.e., cytotoxic concentrations) be used to predict basal toxicity of all cells in the body?
2. Can simple, economic, *in vitro* cell systems be used to measure basal cytotoxicity, i.e., the toxicity of a chemical which can be measured as affecting basic cellular functions and structures common to all human specialized cells, regardless of their organ of origin? Basal cytotoxicity from a chemical, therefore, refers to the measurement of cell membrane integrity, mitochondrial activity, or protein synthesis, since these are examples of basic metabolic functions which are common to all cells. The level of activity, however, may vary from one organ to another depending on its metabolic rate and contribution to homeostatic mechanisms, but the processes do not differ qualitatively among organs. In many cases, target organ toxicity in man seems to be caused by basal cytotoxicity of chemicals distributed to the corresponding organ. If this hypothesis is true, tests using specific cell lines would cover a large percentage of toxic effects and would reduce the need to introduce many laborious systems with specific cells.

Organ-specific primary and continuous cultures have been used to test for systemic toxicity. These cultures are capable of demonstrating basal cytotoxicity, similar to the toxicity shown by undifferentiated cells (such as HeLa cells and Swiss 3T3 mouse fibroblasts), without the need to measure specific functional characteristics. Organ-specific cells also express genuine site-specific cellular toxicity. A chemical capable of altering a particular function which is unique to an organ may raise the question of whether the effect is due to basal or organ-specific toxicity for that agent. A comparison of the dosages of chemicals necessary to produce cytotoxic effects in cell lines of different origin may elucidate the underlying mechanisms of the toxic response.

The use of organ-specific cells to test for systemic toxicity may also be suitable as a screening system, since these cells can provide information on basal and specific cytotoxicity. For example, hepatocytes may be used for the determination of basal cytotoxicity as well as organ-specific toxicity, based on the extensive metabolic capacity of the liver to screen most circulating xenobiotics. A serious disadvantage to the use of hepatocytes, in particular, and many organ-specific primary cultures, in general, is that they cannot be routinely maintained in continuous culture and must be established for each experiment. In addition, primary cultures are more expensive than continuous cell systems and can raise legitimate ethical questions when animals are needed as the source of cells for repeated experiments. Thus, a longterm objective of cytotoxicity testing will be the determination of the role of continuous cultures of cells originally derived from an organ and its relationship to similar cells in primary culture. The investigation of this objective may answer the question, can cells which have been maintained in culture over several passages be used as a reflection of the organ's response to a chemical insult?

II. PRACTICAL APPLICATIONS OF CYTOTOXICITY TESTS

A. APPLICATIONS OF *IN VITRO* TESTS

It is conceivable that a standardized battery of *in vitro* tests can be developed for assessing acute local and systemic toxicity, based upon the validation studies currently underway. These methods and procedures would have many applications.

1. As Part of a Battery of Tests

Certain methods assigned to the battery could be used for systematic mapping of the different effects attributed to basal cytotoxicity. In addition, the mechanisms underlying basal cytotoxic phenomena may explain how these effects are influenced by the physicochemical properties of the molecules. This knowledge is useful for the interpretation of results in cellular testing and can be applied on actions of chemicals known to be cytotoxic to animals or humans by comparing the *in vitro* and *in vivo* concentrations of the same chemicals.

A battery of selected protocols may function as a primary screening tool before the implementation of routine animal testing. The animal tests would then function to confirm the results from the test battery, as well as to detect toxic substances which have eluded the screening protocols. When the refined animal tests have been performed, all available information is used as a basis for the prediction of human toxicity. Thus, the indication of a basal cytotoxic mechanism can be studied using cell line methods. In addition, results from earlier mechanistic studies may be pursued if the mechanism is relevant to chronic toxicity, in which case it would be possible to reduce and refine the animal tests.

With the refinement of general *in vitro/in vivo* testing and the progress afforded such studies with time, certain tests will be employed with greater confidence. This will occur as it becomes apparent that the battery reflects the acute toxicity of certain groups of substances, resulting in the inevitable elimination of certain aspects of animal testing.

2. As Screening Protocols

Today, thousands of relatively common chemicals have never been tested in any system, mainly due to the cost of animal tests. Many new substances are added each year from pharmaceutical and industrial concerns. These chemicals could be assessed for their risks to human health using routine cellular carcinogenicity and cytotoxicity tests.

3. For the Determination of Risk Assessment

Together with genotoxicity tests, *in vitro* cellular test batteries could be applied as biological markers of chemically induced risks, whether man-made or naturally occurring. Such risks include biomonitoring of food and food additives and water and air analysis for environmental toxicants. If similar batteries of protocols are validated for their ability to screen for or predict human toxic effects, these measurements will establish a significant frame of reference for monitoring of environmental and societal threats.

B. APPLICATIONS OF *IN VITRO* TESTS IN COMBINATION WITH *IN VIVO* DATA

It is possible to replace toxicity testing of existing chemicals in the environment with risk assessment based on the results from the battery. This may be combined with sensitive chemical and analytical measurements of human blood and tissue concentrations of the same chemicals, such as with volunteer testing of dermal and ocular irritancy, or through the tabulation of data bases with reliable clinically relevant human toxicity information.

III. FUTURE PERSPECTIVES

Based on the observable trends and practical experience of many laboratories and regulatory agencies with *in vitro* cytotoxicity tests, it is conceivable that toxicity testing in the foreseeable future will be regularly performed with cell cultures and in test tubes. Furthermore, such testing will be more predictive of human toxicity than the current animal tests. If this forecast is realized, many new types of *in vitro* toxicity and toxicokinetic tests will be developed, while many of the present methods will have been validated in refined programs. In addition, some of the validation studies will be based on relevance-focused multicenter studies using human data obtained from available data bases and controlled trials using volunteers. These studies will probably include refined multifactorial modeling of many *in vitro* data to account for human toxicity, including computerized physiologically based kinetic modeling. Simultaneously, the gradual acceptance of new *in vitro* methods should not only depend on formal validation programs, but may be a consequence of the parallel experience of various industrial laboratories using both *in vitro* and *in vivo* methods. Finally, these achievements should not be based only on political or societal attempts to prevent the use of animals in research, but should complement a very efficient and reliable system for protecting the public interest.

INDEX

M

3-Methylcholanthrene, 143

Microdissection, 66

Macrophage activating factor (MAF), 111, 113

Micronization, 176

Micronuclei detection, 153

Macrophage chemotactic factor (MCF), 111, 113

Microscopy, 50, 112, see also specific types

Macrophages, 60, 75, 111–113, see also specific types

Microsomal enzymes, 124, see also specific types

Madin-Darby canine kidney (MDCK) cell lines, 28, 67, 68

Millipore filter method, 89

Mineral oil, 178

MAF, see Macrophage activating factor

Mitochondrial function, 85

Magnesium, 79

Mitotic division, 36

Maintenance cultures, 7–9, 123

Modified Eagle's Medium (MEM), 12

Mammalian mutagenesis, 147–150

Monocytes, 75, 111–113

Manganese, 17

Monolayers, 21

Marijuana, 115

Monooxygenases, 60, 74

Mast cells, 108

Morphine, 113, 115

MCF, see Macrophage chemotactic factor

MTT assays, 37, 53–56, 103

MDCK, see Madin-Darby canine kidney

Multicenter Evaluation for In Vitro Cytotoxicity (MEIC) project, 40, 41, 198, 200–203

Mechanisms of toxicity, 27, 33–41

basal toxicity data and, 37–42

Multilaboratory validation, 197–202

cell line use and, 34–37, 40–41

Muscle cells, 60, 61

lethal dose tests and, 33–34, 37

Mutagenesis, 135, 145–150

organ-specific, 41–42

Mutagenicity, 27, 47–48, 123, see also Mutagens

Mechanistic response to chemicals, 28

Media, 12–19, see also specific types

Mutagens, 127, 135, see also Mutagenicity; specific types

basal Eagle, 12

buffering of, 15–16

Mutations, 27, 40, see also Mutagenicity

chemically defined, 3, 12–17

base-pair substitution, 145

components of, 13

detection of, 127, 133

Dulbecco's modified Eagle's, 12, 13, 67, 70

frameshift substitution, 145

induction of, 128

incubation, 176–178

somatic cell, 129

lymphocyte separation, 109

substitution, 145

separation, 109

Mycoplasma, 17

serum-free, 9, 17–19

Myocardial cells, 29

MEIC, see Multicenter Evaluation for In Vitro Cytotoxicity

MEM, see Modified Eagle's Medium

N

Mesenchymal cells, 2, 7, 28

Metabolic activation systems, 139–144

β-Naphthaflavone, 140

Metabolic cooperation, 148

Natural immunity, 107–108

Metabolic studies in cell cultures, 121–122

Natural killer (NK) cells, 108–111

Metabolism, 42, 50, 59, 60

Neoplastic tissues, 3

enzyme, 123–124

Neural tissue, 63

indicators of, 85

Neuroblastoma cells, 28

liver, 121

Neutral red, 50, 53–54, 91, 103

phase II, 141

Neutrophils, 75, 108, 113–114

products of, 63

Nitroblue tetrazolium, 73

xenobiotic, 60

Nitrofurantoin, 84

Metals, 17, 34, see also specific types

Nitrogen dioxide, 95

Metaphase spreads, 151

Nitroso compounds, 141, see also specific types

Methanol, 53

Methotrexate, 148

NK, see Natural killer

Nonparametric tests, 182, 186–187